AF247444

Rare-Earth Doping
of Advanced Materials
for Photonic Applications–2011

MATERIALS RESEARCH SOCIETY
SYMPOSIUM PROCEEDINGS VOLUME 1342

Rare-Earth Doping of Advanced Materials for Photonic Applications–2011

Symposium held April 25–29, 2011, San Francisco, California, U.S.A.

EDITORS

Volkmar Dierolf
Lehigh University
Bethlehem, Pennsylvania, U.S.A.

Yasufumi Fujiwara
Osaka University
Suita, Osaka, Japan

Tom Gregorkiewicz
University of Amsterdam
Amsterdam, The Netherlands

Wojciech M. Jadwisienczak
Ohio University
Athens, Ohio, U.S.A.

Materials Research Society
Warrendale, Pennsylvania

CAMBRIDGE UNIVERSITY PRESS
Cambridge, New York, Melbourne, Madrid, Cape Town,
Singapore, São Paulo, Delhi, Tokyo, Mexico City

Cambridge University Press
32 Avenue of the Americas, New York, NY 10013-2473, USA

www.cambridge.org
Information on this title: www.cambridge.org/9781605113197

Materials Research Society
506 Keystone Drive, Warrendale, PA 15086, USA
http://www.mrs.org

First published 2012

CODEN: MRSPDH

ISBN: 978-1-60511-319-7 Hardback

CONTENTS

ZnO, GaN, PHOSPHORS

*Invited Paper

PREFACE

Symposium V "Rare Earth Doping of Advanced Materials for Photonic Applications" took place during the 2011 Spring Meeting of the Materials Research Society in San Francisco, California from April 25–29, 2011. The scientists who attended the symposium came from 14 different countries around the world. The conference program had 10 invited talks, 37 oral communications, 20 posters and several short presentations from poster authors. This special symposium proceeding, also available in Online Proceedings Library (OPL) hosted on Cambridge Journals Online (CJO), contains 17 papers, including the invited talks as well as selected oral and poster contributions.

Symposium V successfully fulfilled the organizers aim to bring together researchers from a number of fields that traditionally do not interact closely with each other although their general theme of rare earth doping and using the functionality that these ions offer is common to all of them. The symposium provided to the semi-conductor, phosphors and device communities a unique opportunity to discuss the fundamental topics of common interest that underlie the emission in rare-earth-doped materials. Such a mix of different research topics, silicon photonics, phosphors, oxides, and wide band gap materials including III-nitride semiconductors, to name a few, greatly promotes a healthy and vigorous exchange of ideas.

The goal of this symposium was to highlight the status of light emission at infrared and visible wavelengths from rare-earth-doped phosphors as well as semiconductors. The symposium addressed topics from basic to application-driven research. The overlap of the different areas is very apparent in the discussion about the way rare earth ions can get excited most efficiently and how they can transfer their excitation. Progress has been reported in generating and understanding luminescence from rare earth ions sensitized by nanoclusters in different materials. The demonstration of the efficient electroluminescence in Europium-doped GaN light-emitting diodes at room-temperature has proven the practicality of using this material as a technological contender for optoelectronics. Furthermore, issues of rare earth materials applications for green technologies, sustainability and opportunities for development of multifunctional devices utilizing magnetic, electric and pressure stimuli were addressed in the symposium.

The editors would like to thank the authors of the manuscripts. MRS meetings have become one of the most important forums for rare earth doped materials and applications. The challenges in fundamental issues of generating light in ultraviolet, visible and near infrared spectral regions have a great impact not only on the rare earth ions research community but also on the general fields of photonics and optoelectronics. This volume is a useful resource to share interests within this broad research community.

Volkmar Dierolf

Yasufumi Fujiwara

Tom Gregorkiewicz

Wojciech M. Jadwisienczak

August 2011

MATERIALS RESEARCH SOCIETY SYMPOSIUM PROCEEDINGS

MATERIALS RESEARCH SOCIETY SYMPOSIUM PROCEEDINGS

Prior Materials Research Society Symposium Proceedings available by contacting Materials Research Society

ZnO, GaN, Phosphors

Mater. Res. Soc. Symp. Proc. Vol. 1342 © 2011 Materials Research Society
DOI: 10.1557/opl.2011.1276

Rare Earth Materials Challenge to National Defense: Material Scientist's Perspective

Shiva Hullavarad[1], Nilima Hullavarad, Joclyn Cook[2]
Advanced Materials Group, Institute of Northern Engineering, University of Alaska Fairbanks, Fairbanks, Alaska

Abstract

This article provides the key technical niche that alternative rare earth metals and oxides can offer as an alternative to the ones that are under 'Potential China Export Embargo'. The potential areas of national interest that affected by limited rare earth and the implications of such short supplies on the US businesses are discussed. The paper discusses the technology areas where US based academic and industries have an opportunity in developing the alternate rare earth materials (REM) through innovations in recycling existing rare earth (RE) metals/oxides and develop alternate solutions. Some examples are provided on how the nanotechnology research in the alternative material technologies in the rare earth metals and oxide materials significantly affect the industry trend of rare earth dependence.

Introduction

Rare earth metals and oxides are used to fabricate phosphors to convert the blue light to bright white light, critical optical components for fabricating lasers, high density magnets used in computer hard drives, high k metal dielectrics, and notably in the automotive rechargeable batteries. The US military depends heavily on some of the rare earth metal oxides for strategic applications in missile defense, communications and night vision electronic devices. China mines and exports more than 90% of the rare earths and thus controls the business of all electronics that use the rare earth metal based oxides.

Table 1. North American Rare Earth Reserves (Ref.1. Cox, 2010)

Factor	Mountain Pass (USA) RCF, Goldman Sachs & Traxys	Hoidas Lake (Canada) Great Western Minerals Group	Nechalacho (Canada) Avalon Ventures Ltd	Bear Lodge (USA) Rare Element Resources Ltd
Status	Re-commissioned separation plant. Feasibility study of re-commencing mining and processing underway.	Advanced exploration. Some preliminary test work completed. Could be supplemented by RareCoProject in South Africa	Pre-feasibility study underway. Some preliminary test work completed.	Resource engineering study underway. Process development commenced
Resource	20Mt @9.2% REO 1.8Mt REO contained (a proven reserve)	2½Mt @ 2.4% REO 0.06Mt REO (inferred)	69Mt @2.0%REO 1.3Mt REO (inferred)	9 Mt @ 4.1% REO 0.4 Mt REO (inferred)
Potential Production	Target: 18,000t pa REO; start-up in 2012	3-5,000 tpa REO Start-up post 2014	3-5,000 tpa REO Start-up post 2014	Unknown
Critical Issues	New owners Completing DFS Re starting an 'old' plant.	Define ore reserve Develop process	Define ore reserve Develop process	Define ore reserve Develop process

The U.S. House of Representatives, anticipating that the U.S. military might become dependent on Chinese-made electronics, approved H.R. 6160, the Rare Earths and Critical Materials Revitalization Act of 2010,during the fall of 2010. Table 1 provides the proven North American rare earth reserves[1]. Realizing rare earth importance on critical technologies, the US government has issued grants to find those alternatives or even how to produce our own rare earths[2]. The U.S. government (SBIR/STTR, RFP) invested $400 million to create materials that require smaller amounts of rare–earth metals in permanent

---

[1] Corresponding author: sshullavarad@alaska.edu
[2] Undergraduate student

magnets used in electric motors and wind turbines improve production processes and expand recycling facilities. The USGS recognizes U.S. Rare Earth's deposits as among a select group of economically viable deposits outside of Chinese control. The increased global competition and off shore supply vulnerability of REM that are critical to the emerging clean energy technologies (wind turbines, hybrid motor vehicles, catalytic converters)[3] pose a roadblock in moving towards energy independence from fossil fuels[4].

This article addresses specific applications of REM and its implications to the national defense. Also, we will outline how the rare earth supply vulnerability would affect the national green energy initiative. We will discuss the academia, industry and government laboratory synergistic efforts could provide new avenues for new alternatives to the rare earth, explore untapped rare earth reserves through advanced technologies and recycling existing REM.

Rare Earth – Challenge to Strategic National Defense

In the spring of 2010, U.S. Rep. Mike Coffman, R-Col, introduced a bill mandating the defense secretary and other agencies (commerce, energy, interior) to reexamine the U.S rare earth reserves, industry, and supply/demand aspects leading into creating the nation's key rare earth reserve build up[5]. Department of Defense (DOD) depends heavily on the REM which form the basis of many tactical operations. For example, DOD military platforms - Missile -*Trident, Minuteman IV, Patriot, PAC III, Tomahawk Cruise Missiles, JDAM's, Hellfire missiles, Harpoon Anti-ship*; Navy - *Aegis radar, Virginia Submarines, CVN, DDX, LCS, UUV's, support "all electric ship" technologies, Firefinder radar*; Army – *M1A1 tanks, Bradley A3 and FIST, Paladin Howitzer, AH-64 Apache, Striker Humvee*; Air Force – *F-15, F-16, F-18, B-52, Joint Strike Fighter, Unmanned Aerial Vehicles*; all use the rare earths in some or the form ranging from strategic communications, coatings, and shields. As per the April 2010, U.S. Government and Accountability Office report to the U.S. Senate and U.S. House of Representatives, the DOD's Office of the Director of Industrial Policy ensures the reliable, and cost-efficient rare earth suppliers to meet the demands and also mitigate the short-term supplier base gaps that are critical to the known 24 weapon systems involving precision guided munitions, radar systems, night vision equipment, lasers and satellites[6]. The best possible scenario to address the supply vulnerability requires to follow a multiple pronged approach; (1) identify key rare earth elements that are common to multiple military platforms (lanthanum, samarium, yttrium), (2) develop workforce in the alternative rare earth technologies, (3) form consortium of DOD funded academic research programs.

Alternative to rare earth materials
In our study on the existing REM for specific applications in phosphors, white noise (stealth) and RF areas, the REM are found to have an advantage in their intrinsic 4f-3d electron configurations. REM have similarity in the outer electronic structure – (starting from lanthanum the gradual filling of 4f shell till the configuration is reached to that of lutetium), weakly bound relative to Fermi level but highly localized 4f wavefunctions that are responsible peculiar magnetic properties. Gunnarsson and Schonhammer[7] many-body theory calculations reveal that the dynamic screening by conduction band electrons involving 4f levels (f screening) leads to interesting properties observed in REM. The average atomic size of the rare earth in the matrix of mixture also plays the dominant role in determining the application and sensitivity. Thus the research on new alternative materials for rare earths narrows down to materials fabrication engineering by dynamic non-equilibrium processes like nano/pico second high energy pulsed laser synthesis of materials. The alternatives to REM are discussed below in two specific categories of phosphors and magnets.

Phosphors
Lanthanum oxide is used as phosphor to produce white light to convert typical blue light out of the solid state light emitting diodes. With the drive to replace the incandescent and compact fluorescent lights by

LED's to improve the power efficiency, there will be ever increasing demand for the rare earth phosphor. The new non rare earth phosphors should have the high optical efficiency at lower dopant concentrations, as the higher dopant concentration leads to optical (luminescence) quenching[8]. The novel classes of wavelength conversion materials for white lights will provide the exit off of the rare earth phosphors.

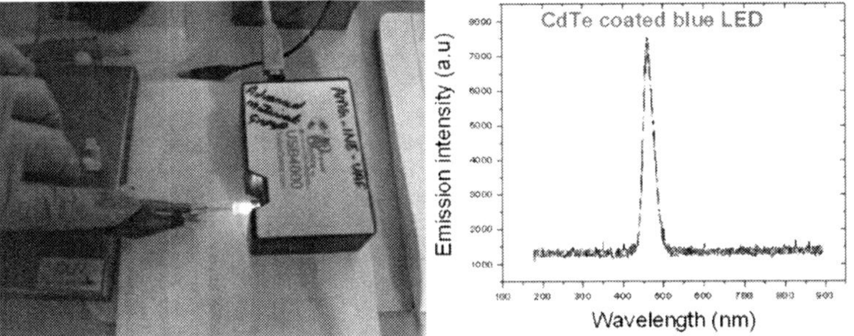

Figure 1. LED coated with nano CdTe and the emission spectrum

To demonstrate the possibility of using the non-rare earth phosphor, we deposited CdTe monodispersed (particle size ~2 nm) in solution form on a commercially obtained blue LED using spin coating method. The coating was allowed to dry and then annealed in nitrogen at 200 C for 30 minutes. The LED was then optically characterized through an Ocean Optics spectrometer for studying the emission once the LED is turned on with 1.5V. The results are shown below in Figure 1. The emission intensity is shifted from the original blue light at 398 nm to the now CdTe coated LED at 482 nm. Though the light wavelength has not shifted much from the original wavelength and the spectrum has not broadened to provide the requisite color balance for the white light (usually a yellow phosphor or a combination of green and red colors would give the white light), the attempt made here demonstrates that it is possible to substitute the rare earth phosphors with suitable choice of coating and most importantly a selection of proven and established industrial manufacturing technique like spin coating, which is scalable for commercial production.

Stealth Technology
The present day laser stealth technology depends heavily on the rare earth oxides. The laser arms consist of two kinds: one is to emit laser to the objects, and then receive the reflective waves, such as laser telemeter, laser-guided apparatus, etc; the other is the laser directly irradiating the targets. So the targets are coated to nullify the laser intensity reflected by the targets in order to make the objects camouflaged. Doped Samarium oxide absorbs the 1.0 -10.5 μm laser thus making an irreplaceable material candidate in the stealth technology[9]. However, infrared-laser- visible multifunctional stealth coatings doped with inorganic semiconductors, whose infrared emissivity was 0.5 - 0.9 at 8-14 μm and laser reflectivity for 1.05 μm was less than 0.4%- which are non rare earth based can be explored as an alternative to samarium oxide.

Superalloys and Magnets
Neodymium is used a rare earth dopants in the concentration range from 12-15% in NdFeB thin films for achieving superior magnetic properties with Br = 12:2 kG, Hc= 9:9 kOe and (BH)max= 35MGOe. The potential use of NdFeB thin film permanent magnet for micro-mechanical systems, micro-electronics and micro-magnetic devices, millisize motors or actuators. By substituting Nd with Lu or Hf in similar

5

concentrations it is possible to achieve similar magnetic properties. Similarly, yttrium in super alloys can be replaced by elements like Cr, or Al to obtain similar materials behavior.

Nanotechnology – A possible route to alternate rare earth engineering at the nanoscale
Nanotechnology offers some unique avenues in the rare earth alternatives (and at the same time reduction of rare earths) research and the eventual transformation of those laboratory developed technologies to the market. One cannot simply replace the rare earths that are in use today with something new, at least in near future or till the new alternative non rare earths are developed completely, tested, adopted and integrated in the mainstream like car engines or wind turbines. So, the science of doing with small, nanoscale engineering makes it possible to provide the high strength magnet functionality using the nanostructured magnets using smaller amounts of rare earth metals like neodymium and dysprosium. These nanostructured regions of neodymium and dysprosium in the iron based nanoparticles magnet interact in a way that leads to greater magnetic properties than those found in conventional magnetic alloys.

The advantage of nanocomposites for magnets is twofold: nanocomposites promise to be stronger than other magnets of similar weight, and they should use less rare-earth metals. The present day advancements in nano scale engineering methods combined with vast knowledge on the spin-exchange coupling mechanisms responsible for high strength magnets is sufficient enough to move towards the new materials research area of non-rare earth nanostructures. As an alternate route, it is possible to fabricate nanoscale magnetic material mixtures like CoPt, Permalloy, Ni, Mn – Oxides and engineering to achieve the materials compositions to reach high-energy product in magnetic matrix through aligned spin-exchange coupling of soft and hard magnetic phases in the range of 30-80 nm by dynamic non-equilibrium processes like nano/pico second high energy pulsed laser synthesis of materials, which can be scaled up for higher throughput.

Table 2. The rare and non – rare earth materials used in fabricating the magnets (Ref.10).

	Compound	Structure	Saturation Magnetization (kG)	Curie temperature (°C)	Anisotropy constant K1 (MJ-m^{-3})
Rare Earth	$Nd_2Fe_{14}B$	Tetragonal	16	312	5
	$SmCo_5$	Hexagonal	11	680	17
Non-Rare Earth	MnAl	Tetragonal	6.5	377	1.7
	MnBi	Hexagonal	8	357	1.2
	$BaFe_{12}O_{19}$	Hexagonal	5	450	0.33
	Zr_2Co_{11}	NA	65 emu/g	500	NA
	$E\text{-}Fe_2O_3$	Orthormbc	14 emu/g	NA	NA
	FePt	Tetragonal	14.5	477	6.6
	Co	Hexagonal	18	1115	0.53
	CoPt	Tetragonal	10	567	4.9
	Co_3Pt	Hexagonal	14	727	2.0

Table 2 provides the list of rare earth and non rare earth magnetic alternative materials[10]. The material(s) system exists in either of two spin states within the thermal hysteresis loop conferring to bistability and retaining the magnetic memory on the nanoscale. The typical requirement for high strength magnets is the large remanence –M, which is a result of alignments of all grains and particles. The present day commercially available magnets are prepared by alloy mixture of Fe, Cu, Zr, Co combined with Sm or Nd powder metallurgy and sintering to obtain the magnets with strength of BH_{max} higher than 25MGOe. Such processes with the rare earths have been successful in driving the magnetic materials industry which forms the backbone of the electromagnetic based components and systems. However, the paradigm of replacing the REM puts the onus back on the intelligent materials engineering that addresses two important material aspects that are critical in the magnets; (1) high anisotropy and (2) proper microstructure. With the selective doping of non-rare earth magnetic elements like Co, Ni, Mn in selective matrix of oxides the desired magnetic properties can be achieved by one of the following mechanisms; (1) nucleation of reversed magnets, (2) domain wall pinning and (3) magnetization rotation.

Academic-Government-Industry synergistic efforts – Tripartite approach
In April 2011, the Department of Energy, Advanced Energy Project Agency announced the REM in critical technology (REACT) focused on two aspects; (1) high strength magnets and (2) high temperature superconductors. The industry has advanced the magnets based applications typically used in electromagnets (samarium, neodymium) which are rare earth based high strength magnets. The research and developments in superconductors area is primarily led by academic institutions in the last two decades, to improve the critical temperatures, Tc, with the hope to achieve the room temperature superconductivity. The search started with Yttrium Beryllium Copper Oxide (also known as YBCO or 123 compound) with Tc at 86K and in last 25 years, the Tc has reached barely 130 K, as per the recent reports on the status of superconductivity[11].

U.S. academic institutions and commercial businesses have led the research and development in utilizing rare earth based compounds for advanced electronic and electromagnetic applications. The following chart provides the synergistic efforts between the academic, industry and government laboratories in spearheading the rare earth research. University of Delaware has a program to develop the rare earth alternative magnets based on transition metal dopants/alloys.

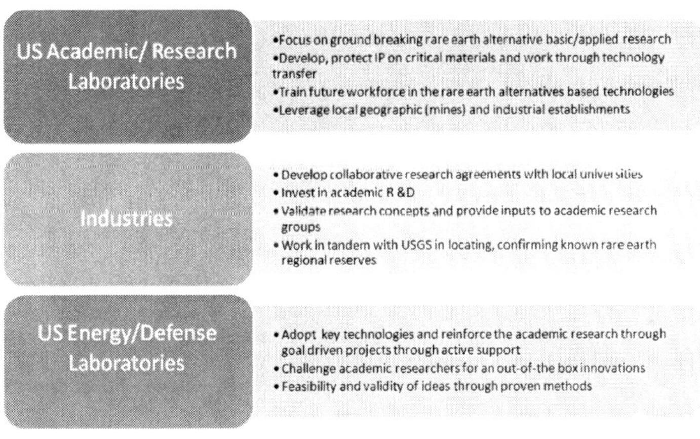

Figure 2. Academy, industry and government synergistic efforts.

The industries have a larger role to play in addressing the rare earth needs and develop alternatives to rare earths, recycling the existing REM and explore untapped rare earth reserves in US. U.S. Rare Earth Inc. recently signed an agreement to explore the large deposits of rare earth metals located (the deposits are also confirmed by USGS) at the Diamond Lake and Lehmi Pass, near Salmon, Idaho and Montana.

Conclusions

U.S. has the technical knowhow and expertise to mine, and process the REM that are spread all across the country in yet to be mined/explored reserves[12]. The new alternative solutions to rare earths in the form of substitutes or smart usage, though promising, but may not be available on a full product till 2015. Besides, the Chinese domestic consumption of rare earths will exceed its own domestic supply soon, opening a new dimension to the already dire situation. This is one opportunity for materials researchers to find the alternative solutions in a reasonable time and transfer to the industrial manufacturing. Academia, industry and government laboratories play an important role in achieving the rare earth independence.

Acknowledgement

The authors acknowledge the financial support from the U.S. Defense Advanced Research Projects Agency (DARPA) at University of Alaska Fairbanks under contract number N66001-08-C-2081.

Disclaimer

The views, opinions, and/or findings contained in this article are those of the author and should not be interpreted as representing the official views or policies, either expressed or implied, of the Defense Advanced Research Projects Agency or the Department of Defense. Approved for Public Release, Distribution Unlimited.

University of Alaska and authors do not, endorse and/or approve any affiliations to any commercial entities and the study reported in this paper is for education and knowledge dissemination purposes only.

References

[1] C. Cox, Industrial Minerals Company of Australia, Rare Earths: Facing New Challenges in the New Decade (2010)

[2] https://arpa-e-foa.energy.gov/#be363c9a-ef01-4b77-956a-6e31b54621cd

[3] Minerals, Critical Minerals and the US Economy, National Research Council, National Academies Press www.nap.edu/catalog/12034.html (2008)

[4] Critical Materials Strategy, US Department of Energy, http://www.energy.gov/news/documents/criticalmaterialsstrategy.pdf (2010)

[5] www.defensenews.com, March 2010

[6] V. B. Grasso, Rare Earth Elements in National Defense: Background, Oversight Issues, and Options for Congress, Congressional Research Service (2010)

[7] O. Gunnarsson and K. Schonhammer, Photoemission from Ce Compounds: Exact Model Calculation in the Limit of Large Degeneracy, Phys. Rev. Lett. 50, 604 (1983)

[8] P.H. Borse, S.K. Kulkarni, N. Deshmukh (Hullavarad), et al., Luminescence Quenching in ZnS Nanoparticles due to Fe and Ni Doping, J. Mater. Sci. 34, 6087 (1999).

[9] L. Huili Laser Absorbency of Samarium Borate Prepared by Solid-state Reaction, J. Rare Earth, 25, 3 (2007)

[10] CRC Handbook of Chemistry and Physics, Ed. David Lide, Edition 79

[11] N.P. Armitage, P. Fournier, R.L. Greene, Progress and perspectives on electron-doped cuprates, Rev. Mod. Phys. 82, 2421–2487 (2010)

[12] K. R. Long, B. S. Van Gosen, N. K. Foley, D. Cordier, The Principal Rare Earth Elements Deposits of the United States—A Summary of Domestic Deposits and a Global Perspective, http://pubs.usgs.gov/sir/2010/5220/

Mater. Res. Soc. Symp. Proc. Vol. 1342 © 2011 Materials Research Society
DOI: 10.1557/opl.2011.994

Electroluminescence Properties of Eu-doped GaN-based Light-emitting Diodes Grown by Organometallic Vapor Phase Epitaxy

Atsushi Nishikawa, Naoki Furukawa, Dong-gun Lee, Kosuke Kawabata, Takanori Matsuno, Yoshikazu Terai and Yasufumi Fujiwara

Division of Materials and Manufacturing Science, Graduate School of Engineering,Osaka Universtiy, 2-1 Yamadaoka, Suita, Osaka 565-0871, Japan.

ABSTRACT

We investigated the electroluminescence (EL) properties of Eu-doped GaN-based light-emitting diodes (LEDs) grown by organometallic vapor phase epitaxy (OMVPE). The thickness of the active layer was varied to increase the light output power. With increasing the active layer thickness, the light output power monotonically increased. The maximum light output power of 50 μW was obtained for an active layer thickness of 900 nm with an injected current of 20 mA, which is the highest value ever reported. The corresponding external quantum efficiency was 0.12%. The applied voltage for the LED operation also increased with the active layer thickness due to an increase in the resistance of the LED. Therefore, in terms of power efficiency, the optimized active layer thickness was around 600 nm. These results indicate that the optimization of the LED structure would effectively improve the luminescence properties.

INTRODUCTION

Eu-doped GaN is expected to realize a GaN-based red LED, which is a key technology for fabrication of the monolithic devices, composed of red, green and blue GaN-based LEDs for full-color display or lighting technology [1, 2]. In order to realize the GaN-based red emission, the fabrication of Eu-doped GaN layers has mainly been performed by ion implantation with post-thermal annealing or molecular-beam epitaxy [3-7]. However, it is rather difficult to obtain a current-injected red emission from a Eu-doped GaN layer fabricated by those methods because of the difficulty in the consecutive growth of high-quality p-n junction structure with the Eu-doped GaN layer. Owing to the successful growth of Eu-doped GaN by OMVPE, we have demonstrated the first current-injected operation of a Eu-doped GaN-based red LED at room temperature (RT) [8, 9]. We have also reported improved luminescence properties with a light output power being as high as 17 μW for the Eu-doped GaN-based red LED grown at atmospheric pressure [10-12]. Although the remarkable improvement was observed in the light output power, it is still insufficient from a practical point of view. The Eu concentrations of the previous studies are under 10^{20} cm^{-3}, where the luminescence intensity still increases with increasing Eu concentration without concentration quenching of the luminescence intensity. Therefore, the number of Eu ions in the active layer should be increased in order to improve the luminescence intensity. It is inevitable to optimize the LED structure as well as to clarify the properties of the Eu-doped GaN active layer. In this study, we investigated the EL properties of Eu-doped GaN-based LEDs with different active layer thicknesses.

EXPERIMENT

The samples were grown on a sapphire (0001) substrate by OMVPE (Taiyo Nippon Sanso SR-2000). The group III and V sources were trimethylgallium, trimethylaluminium and ammonia. Eu ions were doped using tris(dipivaroylmethanate) europium [Eu(DPM)$_3$]. Figure 1 shows the LED structure, which consists of a 20-nm-thick Mg-doped GaN contact layer, a 80-nm-thick Mg-doped GaN layer with a Mg-doping concentration of 5×10^{19} cm^{-3}, a 20-nm-thick Mg-doped Al$_{0.1}$Ga$_{0.9}$N electron-blocking layer, an Eu-doped GaN active layer and a 2.5-μm-thick n-type Si-doped GaN layer with an Si-doping concentration of 5×10^{18} cm^{-3}. The Eu concentration of the active layer was 5.2×10^{19} cm^{-3}, which was estimated by secondary ion-mass spectroscopy for the Eu-doped GaN layer grown under the same growth condition. The active layer thickness was varied from 300 to 900 nm. Thermal annealing was performed in nitrogen ambient at a temperature of 800 °C for 10 min to activate the Mg atoms. Pd/Au and In electrodes were used for ohmic contacts for the *p*-GaN and *n*-GaN layers, respectively. The size of the Pd/Au electrode for the *p*-GaN was 1 mmϕ. The In electrode for the *n*-type layer was deposited on the side of the sample. The EL measurements were carried out with a CDS 610 mini CCD array spectrometer. The light output power of the EL was measured with an integrating sphere spectrometer.

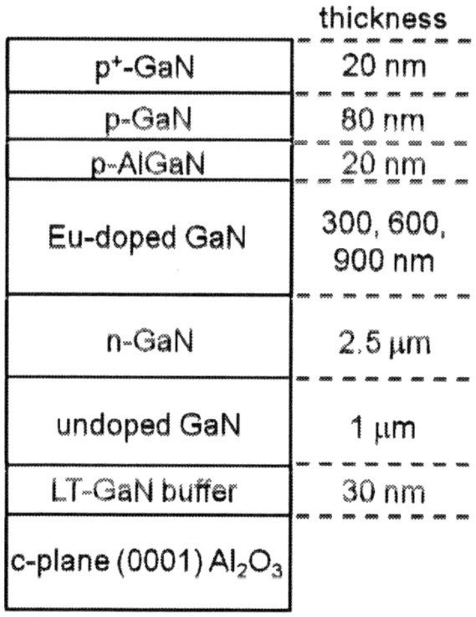

	thickness
p$^+$-GaN	20 nm
p-GaN	80 nm
p-AlGaN	20 nm
Eu-doped GaN	300, 600, 900 nm
n-GaN	2.5 μm
undoped GaN	1 μm
LT-GaN buffer	30 nm
c-plane (0001) Al$_2$O$_3$	

Figure 1. Sample structure of the Eu-doped GaN-based LED.

RESULTS AND DISCUSSION

Figure 2 shows the spectral radiant flux for the LEDs with different active layer thickness, which was measured at RT. The injected current was 20 mA. In the spectrum, emission peaks were observed at around 600, 621, and 663 nm, which are associated with intra-4f shell transitions of $^5D_0-^7F_1$, $^5D_0-^7F_2$, and $^5D_0-^7F_3$, respectively, in Eu ions [13]. The principal peak at 621 nm composed of two peaks and peak intensity ratio changed slightly with the active layer thickness. The increase in the emission peak intensity is attributed to the increase in the number of Eu ions contributing to the emission because of the thicker active layer thickness. The exact linear increase in the peak intensity with active layer thickness was not observed because of the poor hole injection efficiency due to the low activation of Mg atoms in the p-type GaN.

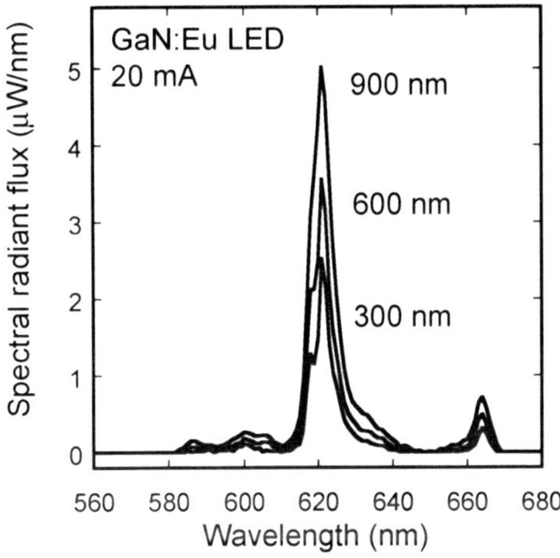

Figure 2. Spectral radiant flux for the LEDs with different active layer thickness.

Figure 3 shows the current-voltage (*I-V*) characteristics of the LEDs with different active layer thickness. The rectifying characteristics were observed in each LED. In the case of the LED with a 300-nm active layer, the current increased remarkably at forward voltages over 3 V, which are close to the diffusion potential of the GaN-based $p-n$ junction. However, the turn-on voltage increased with increasing active layer thickness because of the higher resistance of the LED with thicker active layer.

Figure 4 shows the (a) light output power and (b) external quantum efficiency as a function of injected current for the LEDs with different active layer thickness. The light output power increased with increasing injected current in each LED. The saturation behavior was observed for higher injected current (> 10 mA). Moreover, the reduction of the light output

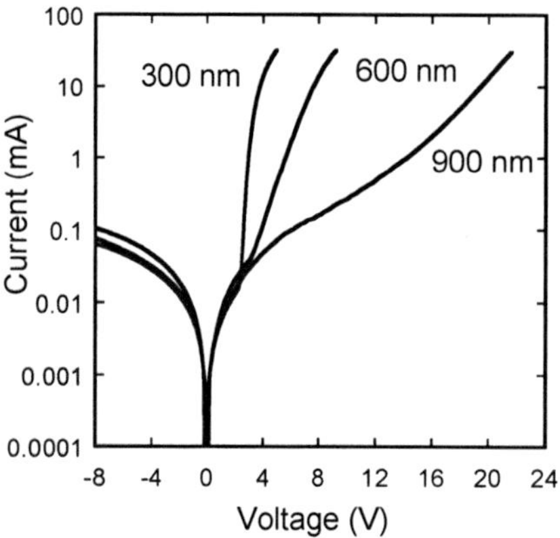

Figure 3. Current-voltage characteristics of the LEDs with different active layer thickness.

power for the LED with a 900-nm active layer was observed. This reduction is presumably due to device heating effect. With increasing the active layer thickness, the light output power monotonically increased. The maximum light output power of 50 μW was obtained for the LED with a 900-nm active layer thickness with an injected current of 20 mA, which is the highest value ever reported. The corresponding external quantum efficiency, ratio of extracted photons to the injected electrons, was 0.12%, which is comparable to that of commercially available GaP:N green LEDs. Because of the saturation behaviour in the light output power, the external quantum

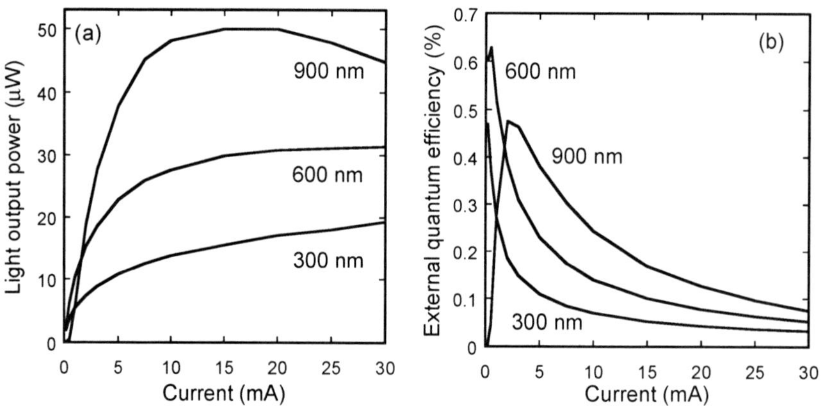

Figure 4. (a) Light output power and (b) external quantum efficiency as a function of injected current for the LEDs with different active layer thickness.

efficiency decreased with increasing injected current. As is shown in Fig.3, the applied voltage for the LED operation also increased with the active layer thickness. Therefore, in terms of power efficiency, light output power divided by the electrical power provided to the device, the optimized active layer thickness was around 600 nm. These results indicate that the optimization of the LED structure would effectively improve the luminescence properties.

CONCLUSIONS

We investigated the EL properties of Eu-doped GaN-based LEDs grown by OMVPE. The thickness of the active layer was varied to increase the light output power. With increasing the active layer thickness, the light output power monotonically increased. The maximum light output power of 50 μW was obtained for an active layer thickness of 900 nm with an injected current of 20 mA, which is the highest value ever reported. The corresponding external quantum efficiency was 0.12%. The applied voltage for the LED operation also increased with the active layer thickness due to an increase in the resistance of the LED. Therefore, in terms of power efficiency, the optimized active layer thickness was around 600 nm. These results indicate that the further optimization of the LED structure would effectively improve the luminescence properties in GaN:Eu LED.

ACKNOWLEDGMENTS

This work was supported, in part, by Grant-in-Aid for Creative Scientific Research No. 19GS1209 from the Japan Society for the Promotion of Science, by Grant-in-Aid for Young Scientists (A) No. 23686099, by the Global Centre of Excellence Program 'Advanced Structural and Functional Materials Design' from the Ministry of Education, Culture, Sports, Science and Technology, and by Murata Science Foundation.

REFERENCES

1. A. J. Steckl, J. C. Heikenfeld, D.-S. Lee, M. J. Garter, C. C. Baker, Y. Wang, and R. Jones, *IEEE J. Sel. Top. Quantum Electron.* **8**, 749 (2002).
2. A. J. Steckl, J. H. Park, and J. M. Zavada, *Mater. Today* **10**, 20 (2007).
3. J. Heikenfeld, M. Garter, D. S. Lee, R. Birkhahn, and A. J. Steckl, Appl. Phys. Lett. **75**, 1189 (1999).
4. S. Morishima, T. Maruyama, M. Tanaka, Y. Masumoto, and K. Akimoto, Phys. Status Solidi A **176**, 113 (1999).
5. H. J. Lozykowski, W. M. Jadwisienczak, J. Han, and I. G. Brown, Appl. Phys. Lett. **77**, 767 (2000).
6. J. H. Park and A. J. Steckl, Appl. Phys. Lett. **85**, 4588 (2004).
7. J. H. Park and A. J. Steckl, J. Appl. Phys. **98**, 056108 (2005).

8. A. Nishikawa, T. Kawasaki, N. Furukawa, Y. Terai, and Y. Fujiwara, *Appl. Phys. Express* **2**, 071004 (2009).

9. A. Nishikawa, T. Kawasaki, N. Furukawa, Y. Terai, and Y. Fujiwara, *Phys. Status Solidi A*, **207,** 1397 (2010).

10. A. Nishikawa, T. Kawasaki, N. Furukawa, Y. Terai, and Y. Fujiwara, *Appl. Phys. Lett.* **97,** 051113 (2010).

11. N. Furukawa, A. Nishikawa, T. Kawasaki, Y. Terai, and Y. Fujiwara, *Phys. Status Solidi A,* **208,** 445 (2011).

12. A. Nishikawa, T. Kawasaki, N. Furukawa, Y. Terai, and Y. Fujiwara, *Opt. Mater.* **33,** 1071 (2011).

13. G. H. Dieke and H. M. Crosswhite, *Appl. Optics* **2,** 675 (1967).

Mater. Res. Soc. Symp. Proc. Vol. 1342 © 2011 Materials Research Society
DOI: 10.1557/opl.2011.1241

Photoluminescence X-ray Excitation Spectra in Eu-doped GaN Grown by Organometallic Vapor Phase Epitaxy

S. Emura[1], K. Higashi[1], A. Itadani[2], H. Torigoe[2], Y. Kuroda[2], A. Nishikawa[3], Y. Fujiwara[3], and H. Asahi[1]

[1]ISIR, Osaka University, 8-1, Mihoga-oka, Ibaraki, Osaka 567-0047, Japan

[2] Graduate School of Natural Science and Technology, Okayama University, 3-1-1 Tsushima, Kita-ku, Okayama 700-8530, Japan

[3] Division of Materials and Manufacturing Science, Osaka University, 2-1 Yamadaoka, Suita, Osaka, 565-0871, Japan

ABSTRACT

X-ray-excited luminescence of GaN doped with Eu ions as a luminescent center was observed in the wavelength range from 350 nm to 650 nm. Three peaks at 375 nm, 550 nm and 622 nm were found. To survey the mechanism of the photoluminescence due to non-resonance excitation, photoluminescence X-ray excitation spectra are also measured. The mechanism of the luminescence occurrence was briefly discussed based on the model developed by Emura *et al.*

INTRODUCTION

As well known, rare earth compounds and rare earth ions as the dopant in some matrices are of uniqueness in the magnetism and the luminescence natures. The uniqueness conducts to many practical devices. In such situation, there are still few examples that rare earth ions as dopant in a semiconductor were utilized for the effective devices. A GaN-based red light-emitting diode (LED) is expected to enable the fabrication of nitride-based monolithic optical devices for full-color displays and/or lighting technology. Recently, Fujiwara and co-workers successfully grew the Eu-doped GaN (GaN:Eu) by organometallic vapor phase epitaxy and reported the first demonstration of the current-injected red emission at room temperature using the GaN:Eu as an active layer of the LED structure [1].

For the improved performance of the LED device, understanding of the local structures around the optically active Eu^{3+} ions should be required. X-ray absorption fine structures (XAFS) method has several advantages. In particular, as the initial states of the electron transition by X-ray irradiation are deep core levels such as a 1s orbital for the rather light elements and a 2p orbital for heavy elements, element selective detection can be carried out.

In this contribution, we try unique XAFS method in the secondary process detecting mode like fluorescence mode and total electron yield method to investigate the local structures around the optically active Eu^{3+} ions. In this method, we do not detect fluorescence X-rays, which is usually adopted for dilute systems, but catch the visible photoluminescence. Even if the several local surroundings around the dopant elements simultaneously exist in the matrix, this method has the potential to distinguish those, because the photoluminescence peak energy is sensitive about the local alignment of the dopant element.

EXPERIMENT

The GaN:Eu samples were prepared by organometallic vapor-phase epitaxy techniques. The group III and V sources were trimethylgallium, and trimethylaluminium and ammonia. Eu^{3+} ions were doped using tris(dipivaroylmethanate)-europium [Eu(DPM)$_3$]. A 3.4-μm-thick GaN/sapphire (0001) template substrate was used. The Eu concentration was determined to be 7 $\times 10^{19}$ cm^{-3} by a secondary ion-mass spectroscopy. The detailed growth parameters and the growth procedure are described in Ref. 1.

The PF ring in KEK (Tsukuba) was selected to record the X-ray excited optical luminescence (XEOL) spectra and the photoluminescence X-ray excitation (PLXE) spectra. The storage ring was operated in a top-up mode with a positron energy of 2.5 GeV and the ring current of 450 mA. The apparatus set for the XEOL measurements was arranged in the experimental hatch of the BL9A beamline. This beamline is installed in a bending magnet port of the storage ring. The synchrotron radiation source beam from the storage ring was monochromatized by a Si (111) double-crystal monochromater, and conducted into an experimental hatch. A focusing mirror is inserted between the monochromater and a final slit, which fixed the form of the beam on the sample. The mirror works as an apparatus of the higher harmonics rejecter. The beam size on the sample was 2×1 mm^2. The energy of the incident X-rays was calibrated at a pre-edge peak of Cu metal. Figure 1 shows a set-up diagram of the observation of XEOL-PLXE spectra. The intensity of the incident X-rays was monitored by a front ionization chamber filled by nitrogen gas. The luminescence from the sample was focused on the entrance slit (1 mm) of a grating monochromater (Jobin Yvon H-20), and analyzed with the exit slit size of 1mm. A photomultiplier (Hamamatsu R5929) was used as the detector of the luminescence. On recording the XEOL spectra, the incident X-rays was hold at the selected energy, and the analyzing monochromater was scanned. The PLXE measurements were carried out at the individual luminescence peaks by scanning the Si (111) double-crystal monochromater to sweep the energy of the incident X-rays. Any filters is not inserted in the path from the sample to the monochromater. All spectra were taken at room temperature.

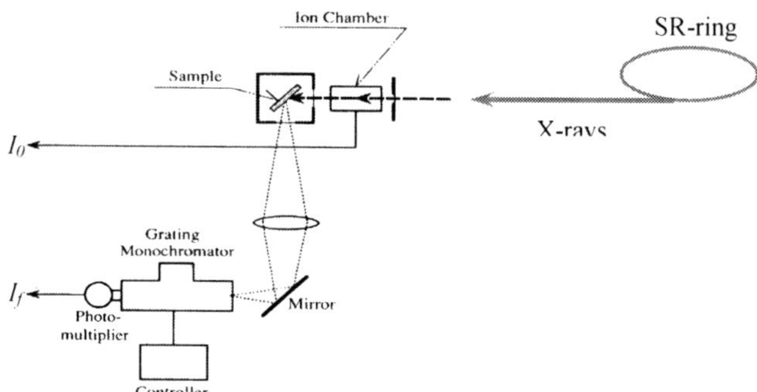

Figure 1. A schematic diagram of experimental setting for XEOL and PLXE measurements. The sample was mounted on the small stage in air.

RESULTS

X-ray excitation draws inner shell transitions, and also creates photoluminescence in the ultraviolet, visible and infrared energy regions as well as X-ray fluorescence (characteristic X-rays). Figure 2 shows the XEOL spectra of the sample GaN:Eu at room temperature. The sample was excited at the higher energies than Eu L_{III}-edge (6977 eV). Three distinguishable peaks are recognized. The sharp line at 622 nm can be attributed to the intra-$4f$ shell transitions in the doped Eu^{3+} ions, and is assigned to the transition from a $^5D_0$ state to a $^7F_2$ state. A broad band around 550 nm, which is frequently observed in GaN, is so-called yellow luminescence. The energy positions of the both bands are completely the same as those of the photoluminescence recorded by excitation of a He-Cd laser (325 nm). The origin of the peak at 375 nm is unclear at present. The spectrum shape suggests a medley of a few photoluminescence peaks. It appears that no proper luminescence of GaN or exciton luminescence is observed.

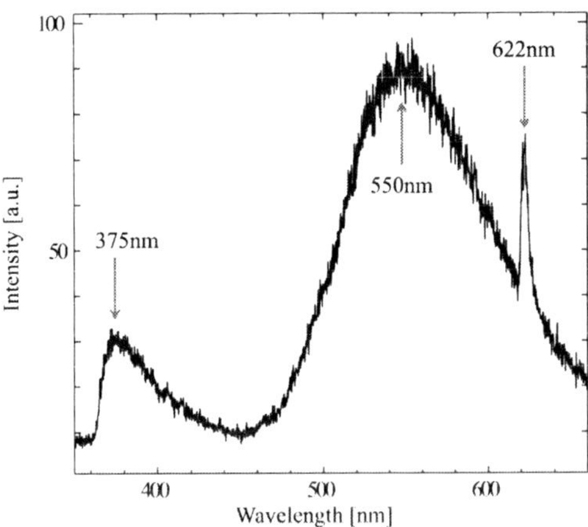

Figure 2. Photoluminescence excited at 7100 eV above Eu L_{III}-edge (6977 eV). A sharp line at 622 nm is assigned to the transition from a $^5D_0$ state to a $^7F_2$ state of Eu^{+3} ion in GaN. The broad band around 550 nm is as the well-known yellow band.

At each photoluminescence peak energy, we tried to observe the PLXE spectra. At first, the Eu^{+3} ion originated peak was examined by fixing the wavelength of the grating monochromater at 622 nm. An edge-jump could not be found around Eu L_{III}-edge (6977 eV). However, the definite edge-jump at Ga K-edge was observed as shown in figure 3. The spectrum entirely corresponded to that observed by the X-ray fluorescence mode (not shown here). It implies that the PLXE spectra observed here give the same information about the local structures as the standard absorption mode. For the yellow band at 550 nm, the PLXE spectrum that resembles closely to figure 3 was recorded. On the other hand, the PLXE spectrum for the UV band at 375 nm showed the inverted order, which is given in figure 4. It is recognized that the XAFS oscillation part is completely reversed. This finding indicates that the mechanism of the photoluminescence occurrence by X-ray excitation is different between the both cases.

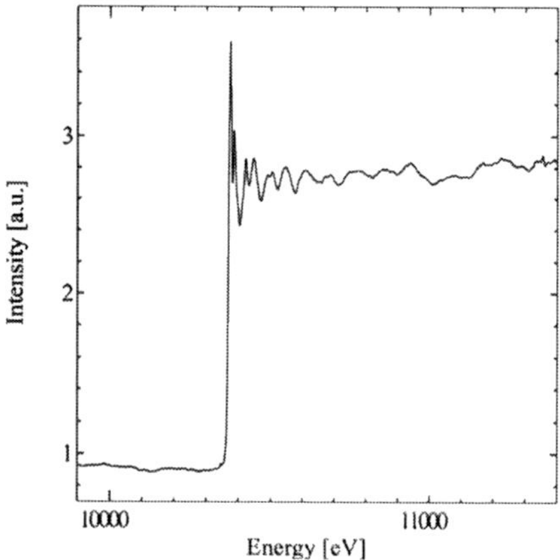

Figure 3. PLXE spectrum around Ga-K edge received at XEOL λ=622 nm.

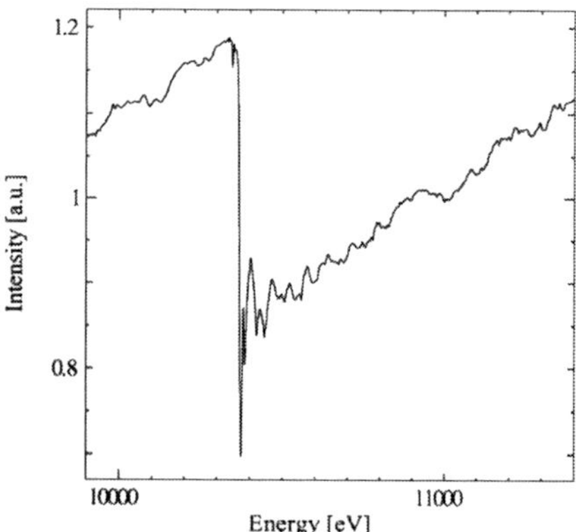

Figure 4. PLXE spectrum around Ga-K edge received at XEOL λ=373 nm.

DISCUSSION

The X-ray irradiation to materials causes various physical events. Photoluminescence is one of them. X-ray excitation is not a resonance process for the photoluminescence occurrence. There are several channels to draw luminescence phenomenon by hard X-ray excitation. At a glance, there are two processes in the mechanisms, the case (A) of the practice rule by hole generation and the case (B) hitting individual ions by energetic electrons (inelastic scattering). These are schematically drawn in figure 5, where three different electron excitations, a relaxation event through a X-ray fluorescence, a *KLL* Auger effect, and inelastic multiscatterings of photo-excited electrons are depicted. Cascade Auger process generates the holes in the steady state like valence band in semiconductors. The excited energetic electrons create many secondary electron and hole pairs or excite the ground state of the dopant ions to a higher state in a process similar with the cathodluminescence.

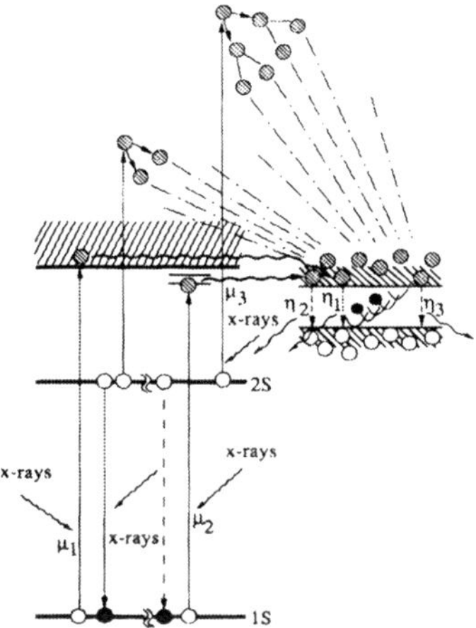

Figure 5. A schematic model for excitation-photoluminescence cycles. Three different excitation processes give rise to an identical photoluminescence with the respective photoluminescence efficiencies η_i. (After Ref. [2])

As described in "RESUTS", we could not find an edge-jump at the Eu L_{III}-edge (6977 eV). This fact implies no or very low contribution to the photoluminescence event from the case (A). It should be noticed here that the incident X-rays with around Eu L_{III}-edge (6977 eV) energy brings the transition from 3s (1800 eV) and $2p_{3/2}$ (6977 eV) orbital in the Eu ion, and the transition of 2s (1299 eV), $2p_{3/2}$ (1116.4 eV) and $2p_{1/2}$ (1143.2 eV) electrons of the Ga ion is coincident, and that the photoelectrons excited from the 2p orbital of the Ga ion have much higher kinetic energy than those from 3s and 2p orbital in the Eu ion [3]. If the case (A) gives no contribution, the photoluminescence observed in this work originates only from the stimulation of the ground state, $^7F_0$, of Eu^{+3} ion by the energetic photoelectrons excited from the 1s orbital of the Ga ion adjacent the Eu ion in major part. The cascade Auger process essentially generates the hole in the steady state of the Eu ion, and may lead to emit the luminescence. As a result, we can find the edge-jump because the hole generation in the steady state by cascade Auger effect is stronger in the deeper core levels.

The counter spectrum given in figure 4 to the standard one (figure 3) is phenomenologically explained, following the model developed by Emura *et al.*[2] The kinetic energy of the photoelectrons created by the excitation of 2s orbital is much higher than those from 1s orbital at the Ga K-edge, resulting in much more secondary electron and hole pair generation than the later excitation from 1s orbital. Scanning the X-ray energy across the Ga K-edge, the spectrum with the inverted order will be observed. By the way, the case (A) brings the standard spectra.

SUMMARY

The photoluminescence by hard X-ray excitation is observed in GaN doped with Eu ions. Three photoluminescence bands are found at 622 nm (intra-$4f$ shell transition of Eu^{3+}), 550 nm (yellow band) and 375 nm (unknown) in the wavelength region from 350 nm to 650 nm. The photoluminescence excitation spectra in the hard X-ray energy region are also recorded to survey the mechanism of the photoluminescence and the local alignment around the dopant ion. No edge-jump at the Eu L_{III}-edge is found, but the clear edge-jump at the Ga K-edge is recognized in receiving the 622 nm and 550 nm bands. On the other hand, the counter spectrum to those for 622 nm and 55 nm is obtained in receiving the 375 nm band. The photoluminescence mechanism of each band is roughly described. No edge-jump at the Eu L_{III}-edge suggests that the photoluminescence of the 622 nm peak is originated in the electron excitation within $4f$ shell of the Eu ion by inelastic scattering of the energetic photoelectron with the Eu ions. The counter PLXE spectrum, which is recorded in receiving the 375 luminescence band, is explained based on the model developed by Emura *et al.*

ACKNOWLEDGMENTS

This work is supported, in part, by a Grant-in- Aid for Creative Science Research No. 19GS1209 from the Japan Society for the Promotion of Science. XEOL and XEOL-PLXE spectra were measured at beamline BL9A of Photon Factory in KEK in cooperation with Institute of Material Structure Science through Proposal 2010G643.

REFERENCES

1. A. Nishikawa, T. Kawasaki, N. Furukawa, Yoshikazu Terai, and Y. Fujiwara, *Appl. Phys. Exp.*, **2**, 071004 (2009).
2. S. Emura, T. Moriga, J. Takizawa, M. Nomura, K. R. Bauchspiess, T. Murata, K. Harada, and H. Maeda, *Phys. Rev.*, **B47**, 6918 (1993).
3. Other outer shells of each element, of coarse, are excited. However, the absorption coefficients for these orbital are very low at 7100 eV.

Mater. Res. Soc. Symp. Proc. Vol. 1342 © 2011 Materials Research Society
DOI: 10.1557/opl.2011.1049

Nature and Excitation Mechanism of the Emission-dominating Minority Eu-center in GaN Grown by Organometallic Vapor-phase Epitaxy

Jonathan Poplawsky[1], Nathaniel Woodward[1], Atsushi Nishikawa[2], Yasufumi Fujiwara[2], and Volkmar Dierolf[1]

[1]Physics, Lehigh University, Bethlehem, Pennsylvania, U.S.A.

[2]Division of Materials and Manufacturing Science, Osaka University, Osaka, Japan.

ABSTRACT

In-situ doped Eu ions in GaN grown by Organometallic Vapor-phase Epitaxy (OMVPE) at different pressures were investigated under different excitation methods and through the use of the following experimental techniques: (1) resonant site-selective laser irradiation (2) electron beam excitation, and (3) a dual excitation using a combination of electron beam and laser irradiation. With these means, we have examined the difference in the excitation pathways that result from resonant laser and electron hole (e-h) pair excitation of Eu ions for two different distinct incorporation sites, which are responsible for most of the luminescence. We have obtained clear evidence that e-h pairs do not have the ability to excite all of the ions and that there is excitation trapping by defects involved in the Eu excitation.

INTRODUCTION

The RE doped GaN materials system has shown to be a promising candidate for LED's, lasers, and displays [1-7]. RE's have the ability to emit sharp spectral lines regardless of the host due to their shielded partially filled $4f^n$ shell, while the wide band gap of GaN decreases the thermal quenching of the ions that correlate to a decreased emission at room temperature [8]. In particular, europium doped GaN is a candidate for the active layer used in a white LED structure to produce the red component of the white spectrum in order to have a pure semiconductor LED instead of one that uses phosphors to down convert. A problem that has been identified is that the emission intensity from excitation by e-h pair recombination is not very efficient and does not scale with the number of Eu centers incorporated into the sample [9]. This suggests that not all of the Eu ions are effectively excited by e-h pair recombination and hence do not contribute to the emission. In order to further improve performance, the excitation pathways and the nature of the most efficient Eu center type have to be better understood. To this end, we use simultaneous excitation of the ions by a visible laser and by an electron beam to study the relationship between resonant direct excitation of the ion itself and excitation of the ion after the creation of electron-hole pairs.

EXPERIMENT

The two samples used in this experiment were grown by OMVPE at 1050° C under two different pressures (10 kPa and 100 kPa) and have a Eu concentration of 7×10^{19} cm^{-3} and 3×10^{19}

cm^{-3} respectively. For more information on the growth of these samples refer to [3,7]. It has been reported that active layers grown under high-pressure (HP) exhibit the best performance in electroluminescence (EL) devices despite the lower concentration and emission spectra that exhibit more disorder and line broadening [7]. It is one of the purposes of this work to explain this behavior.

We have created a new experimental setup that combines the properties of photoluminescence (PL) and cathodoluminescence (CL) by overlapping a laser and an electron beam on the same spot of the sample enabling a wide variety of studies. In order to ensure a good spatial overlap, the collection area is limited to a 2 µm spot by utilizing the techniques of confocal microscopy. Figure 1 shows the experimental setup with an excitation fiber, a collection fiber, and an objective to focus the laser to the same spot from which the emission light is collected. This setup was custom designed to fit on an Oxford Helium Low Temperature Stage that is inserted into a JEOL 6400 scanning electron microscope (SEM) currently allowing T-dependent measurements down to 100K. For e-beam excitation, we used a 10kV acceleration voltage and defocused the electron beam to ensure a homogenous excitation of the entire relevant collection volume on the 300nm thick film (i.e. π x $(1\mu m)^2$ x 300nm). This way the influence of the surface is also minimized.

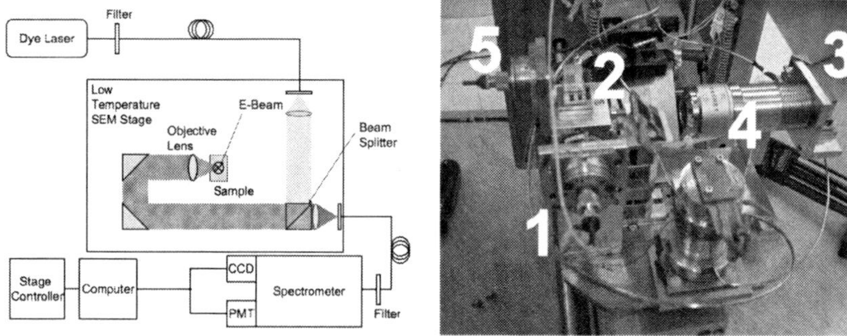

Figure 1. Schematic diagram of the experimental setup (left). On the right is the setup as it is prepared to be inserted into the SEM. The laser enters through the excitation fiber (1), deflects off of the beam splitter (2), reflects off of two 45° mirrors (3), and is focused onto a sample tilted at a 45° angle by an objective (4). The emission follows the same path backward through the beam splitter and into (5) the collection fiber.

The set-up allows us to perform regular photoluminescence excitation spectroscopy (CEES) by using a tunable laser and recording the emission spectra for consecutively changing excitation wavelengths. The resulting dataset is depicted as image plots.

DISCUSSION

Eu sites in GaN

Figure 2 compares, for the two samples, the CEES data obtained for excitation wavelengths in the visible between 570 nm (~2.176 eV) and 573 nm (~2.166 eV) in a spectral region in which a phonon-assisted excitation transition of the Eu ions are found. This excitation method directly excites the $^7F_0$ level of the Eu ions without the formation of e-h pairs. We used the phonon assisted excitation because the emission gives the most reliable data for relative numbers of centers [10], [11]. Woodward et al [10] have revealed 7 different Eu centers in GaN labeled OMVPE 1..7. For e-h recombination excitation, the sites OMVPE 4 and 7 (the centers this paper investigates) dominate the emission and will be labeled Eu1 and Eu2 to better correlate with O'Donnell et al [12].

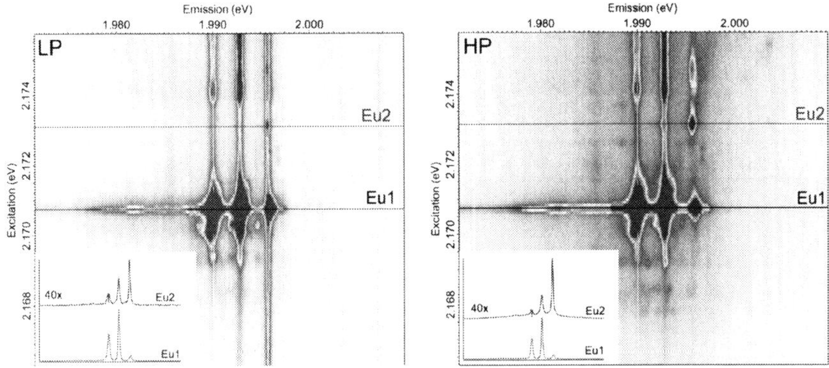

Figure 2. CEES maps for the LP and HP samples that are used to identify different Eu incorporation sites. The horizontal line profiles of the Eu1 and Eu2 sites are depicted in the insets.

In contrast to results from above band gap excitation [13], the resonant excitation experiments show that the Eu1 center, which dominates the luminescence (>97%) under these excitation conditions, scales according to the total number of incorporated Eu ions. Therefore, we assume that it is the center that has been identified as a rather unperturbed Eu on Ga site by Rutherford backscattering studies [13]. We further find that for the HP sample, with a lower overall Eu concentration, the Eu2 center exhibits a more pronounced peak compared to the LP sample indicating a higher concentration of the Eu2 center. This is one reason why it exhibits better electroluminescence behavior. The HP sample shows additional sites and a more pronounced fluorescence narrowing effect indicating lower crystalline quality, additional defects, and a less homogeneous strain distribution. However, the vertical line at 1.996 eV is much less defined. This feature is connected with the ability to excite the Eu2 site non resonantly in the visible and is related to a deep trap. Reduction of this trap further enhances the EL performance.

Overall, it appears that while the total number of defects may have been increased for the HP sample, the types of defects in the HP samples favor EL device luminescence.

Cathodoluminescence and E-H Pair Excitation of Eu1 and Eu2

To further elucidate the role of defects for the excitation of the two centers, we performed excitation power dependant CL experiments with a defocused e-beam in spot mode and compared the saturation behavior of the Eu1 and Eu2 centers (observed at 1.990 and 1.996 eV respectively). Figure 3 shows the spectra associated with a low (0.002 $nA/\mu m^3$) and high (0.4 $nA/\mu m^3$) power density as well as the comparison of the saturation behavior between these samples. Under low excitation densities the Eu2 site dominates in emission, despite its lower abundance, suggesting a more favorable excitation mechanism compared to the Eu1 center.

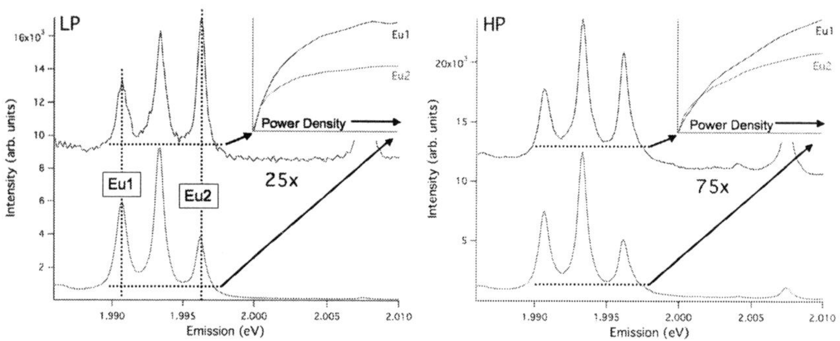

Figure 3. The CL emission of the Eu ions at low (top) and high (bottom) excitation densities. The insets depict the Eu1 and Eu2 peak heights vs. the excitation density, which follows from a minimum value of 0 $nA/\mu m^3$, linearly, to a maximum value of 0.4 $nA/\mu m^3$.

This is consistent with N. Woodward et al. [9] in which even lower excitation power densities were used that show a more exaggerated version of this effect. Such a drastic difference in excitation efficiency between the Eu1 and Eu2 centers is not expected for direct impact excitation and leads us to the conclusion that Eu excitation is dominated by a channel that involves an e-h pair. For higher power densities the Eu1 center begins to dominate due to a saturation of Eu2.

Another important aspect of this data is the comparison of the saturation behavior between the HP and LP samples shown in the insets of Figure 3. Under the highest excitation density used in this study, the LP sample saturates while the HP sample's emission is still increasing. This suggests that in the HP sample, the e-h pairs are able to excite more Eu centers possibly due to the higher number of mediating defects.

<u>**Combined Excitation and the Comparison of E-beam and Laser Irradiation**</u>

In order to study the effect of visible light on the excitation efficiency by e-h pairs, we compared the effects of a resonant (2.171 eV) and non-resonant (2.168 eV) laser excitation under simultaneous e-h pair formation by the e-beam irradiation. Figure 4 shows, for the two samples, the emission intensities observed using just the e-beam, the laser by itself, and a combination of the e-beam and laser. To ensure reliability, the measurements were performed on the identical spot within 10s of each other.

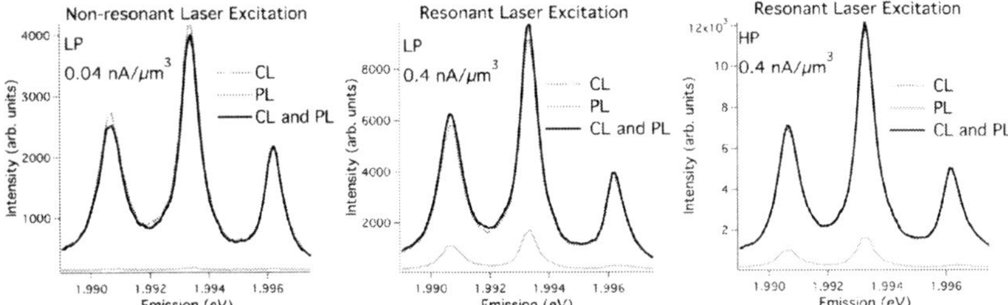

Figure 4. (left) The effect of a laser excitation on the CL emission within the 568-590 nm (2.104-2.185 eV) range used in this study. This was done in combination with an under saturated CL excitation. The resonant excitation of Eu1 under CL saturation conditions for the LP sample (middle) and HP sample (right).

Under non-resonant laser excitation conditions, the CL signal is decreased suggesting that an intermediate trapping process plays a role in the excitation of the Eu ion. Another noticeable effect is that the e-h process of exciting Eu1 is more drastically affected by the laser irradiation compared to that of the Eu2. This suggests that for the Eu2 center the excitation channel is more direct and does not rely as much on intermediate traps. Similar results were observed for the HP sample.

In order to obtain an estimate on how many of the incorporated Eu ions can be excited through e-h pairs, we used a resonant laser excitation of the Eu1 center under an e-beam irradiation with a power density that saturated the CL signal. Although, under these conditions, all Eu ions that can be excited through e-h pairs are excited, a substantial number of Eu ions are still available for resonant excitation for the LP sample, which is evident from the increase in luminescence. For the HP sample, this effect is much smaller suggesting again that in this sample a larger fraction of the Eu ions can be excited through e-h pairs.

CONCLUSIONS

With the goal to enhance the performance of a red LED based on Eu:GaN, we studied two samples grown by OMVPE under different pressure conditions to explain their different electroluminescence properties on the basis of the number of two main center types and their respective excitation pathways. For both samples, we find in PL studies with direct excitation of the Eu ions, several different incorporation sites of the Eu ion in GaN. The Eu1 is clearly dominating in numbers (< 97%). However, the HP sample, which produces better performance in EL devices, shows a higher relative number of Eu2 centers compared to the LP sample. This center has a much more efficient excitation pathway as indicated by its saturation behavior. Investigating the excitation pathways of the Eu1 and Eu2 center by simultaneous excitation in the visible and by an electron beam show a defect mediated excitation channel for the Eu1 center which is unable to excite all of this type of center and, for the Eu2 center, a more direct excitation which by itself may act as a trap for the e-h pair possibly due to an associated nearby defect. Overall, our results indicate that removing all defects in the sample may not be the most ideal situation for the excitation of the Eu ion. Certain defects may enable an effective energy transfer from the host to the ion.

REFERENCES

1. C. Munasinghe, A. Steckl, E. E. Nyein, U. Hommerich, H. Peng, H. Everitt, Z. Fleischman, V. Dierolf, and J. Zavada, Mater. Res. Soc. Symp. Proc. **866**, V3.1 (2005).
2. C. Munasinghe, J. Heikenfeld, R. Dorey, R. Whatmore, J. P. Bender, J. F. Wager, and A. J. Steckl, IEEE Trans. Electron Devices **52**, 194 (2005).
3. A. Nishikawa, T. Kawasaki, N. Furukawa, Y. Terai, and Y. Fujiwara, Appl. Phys. Express **2**, 071004 (2009).
4. C. Ugolini, J. Lin, H. Jiang, and J. Zavada, Appl. Phys. Lett. **97**, 141109 (2010).
5. R. Dahal, C. Ugolini, J. Y. Lin, H. X. Jiang, and J. M. Zavada, Appl. Phys. Lett. **93**, 033502 (2008).
6. J. H. Park and A. J. Steckl, Opt. Mater. **28**, 859 (2006).
7. A. Nishikawa, N. Furukawa, T. Kawasaki, Y. Terai, and Y. Fujiwara, Appl. Phys. Lett. **97**, 051113 (2010).
8. A. J. Steckl, J. C. Heikenfeld, D. S, Lee, M. J. Garter, C. C. Baker, Y. Wang, and R. Jones, IEEE J. Sel. Top. Quant. **8**, 4 (2002).
9. N. Woodward, J. Poplawsky, B. Mitchell, A. Nishikawa, Y. Fujiwara, and V. Dierolf, Appl. Phys. Lett. **98**, 011102 (2011).
10. N. Woodward, A. Nishikawa, Y. Fujiwara, and V. Dierolf, Opt. Mater. **33**, 7 (2010).
11. Z. Fleischman, C. Munasinghe, A. J. Steckl, A. Wakahara, J. Zavada, and V. Dierolf, Appl. Phys. B: Lasers Opt. **97**, 607 (2009).
12. I. S. Roqan, K. P. O'Donnell, R. W. Martin, P. R. Edwards, S. F. Song, A. Vantomme, K. Lorenz, E. Alves, and M. Bockowski, Phys. Rev. B **81**, 085209 (2010).
13. K. Lorenz, E. Alves, S. Roqan, K. P. O'Donnell, A. Nishikawa, Y. Fujiwara, and M. Boćkowski, Appl. Phys. Lett. **97**, 111911 (2010).

Mater. Res. Soc. Symp. Proc. Vol. 1342 © 2011 Materials Research Society
DOI: 10.1557/opl.2011.995

Damage formation in GaN under medium energy range implantation of rare earth ions: a combined TEM, XRD and RBS/C investigation

B. Lacroix[1], S. Leclerc[1], P. Ruterana[1], A. Declémy[2], S.M.C. Miranda[3], K. Lorenz[3], and E.Alves[3]
[1]CIMAP, UMR 6252 CNRS-ENSICAEN-CEA-UCBN, 6 Bd Maréchal Juin, 14050 Caen, France
[2]Institut P', CNRS-Université de Poitiers-ENSMA, SP2MI-BP 30179, Bd Marie et Pierre Curie, 86962 Chasseneuil-Futuroscope cedex, France
[3]Instituto Tecnológico e Nuclear, Estrada Nacional 10, 2686-953 Sacavém, Portugal

ABSTRACT

In this work, the damage formation subsequent to Eu implantation at 300 keV has been investigated by coupling the TEM, XRD and RBS/C techniques. It has been found that GaN exhibits a specific damage buildup in three main steps: (i) clustering of point defects and formation of a network of stacking faults defects in the bulk, (ii) propagation of the planar defect network towards the surface and (iii) breakdown of the surface layer. This occurs through different strain saturation regimes. Around 5×10^{14} Eu/cm^2, the strain along the implantation direction saturates to 0.6%. At higher fluence, whereas the peak at 0.6% is maintained, there is an increase of the strain throughout the implanted layer which probably continues to extend. A second saturation occurs when the stacking fault network reaches the layer surface.

INTRODUCTION

For the ten past years, rare earths (REs) doping of gallium nitride (GaN) has received a great interest due to expected promising applications in the field of optoelectronics and photonics [1,2]. For the fabrication of optoelectronic and microelectronic devices, ion implantation is a conventional and attractive technique to introduce a variety of elements, in a selective area, with a good control of their profiles (concentration, depth). Moreover, such technique avoids the problem of solubility limit, which is the most significant drawback for in-situ doping during the epitaxial growth, as used currently for the fabrication of nitride based devices. Unfortunately, ion implantation also creates structural damage which needs to be annealed out in order to recover the desired electrical or optical properties of the material. It is therefore of prime importance to have a good understanding of the damage formation during the process. Several recent studies have already been devoted to the investigation of the structural evolution in GaN under keV ion bombardment, and it was shown that GaN exhibits a non-conventional behavior in comparison to other semiconductors [3,4]. In particular, it was shown recently that the damage buildup subsequent to Eu implantation proceeds through the formation of a network of basal [5] and prismatic [6] stacking faults which starts to form around the ion mean projected range (Rp); it then extends from the bulk to the surface with the increased fluence, leading eventually to a mechanical breakdown of the surface layer [7]. In this work, we have investigated the effects of europium implantation at 300 keV and room temperature (RT) into GaN. To understand the physical origin of the 'propagation' of the extended defects towards the surface, we present here a correlated investigation using transmission electron microscopy (TEM), Rutherford backscattering spectrometry in channeling configuration (RBS/C) and X-ray diffraction (XRD).

EXPERIMENTS

Gallium nitride (GaN) layers of 2 μm-thick were grown along the [0001] axis on (0001) sapphire substrate by metal organic chemical vapor deposition (MOCVD). Europium (Eu) ions were then implanted into GaN at room temperature, in channeling conditions, with energy of 300 keV and different fluences ranging from 5×10^{13} to 5×10^{15} Eu/cm^2. According to SRIM calculations [8], Rp, the straggling (ΔRp) and the position of the maximum of damage below the surface are 55 nm, 46 nm, and 35 nm, respectively. Structural investigations of the samples were conducted by transmission electron microscopy (TEM), Rutherford backscattering spectrometry in channeling configuration (RBS/C) and X-ray diffraction (XRD). Cross-sections for TEM were thinned down to less than 10 μm by mechanical polishing using the tripod method. The electron transparency was achieved at liquid nitrogen temperature by ion milling at 5 kV using the GATAN precision ion polisher system (PIPS) at an incidence angle of 5 degrees. Conventional and high resolution TEM observations were carried out in JEOL 2010 and JEOL 2010F microscopes operating at 200 keV. The level of damage created during ion implantation was determined by RBS/C experiments, with a 2 MeV He ion beam [9]. XRD measurements were performed on an automated laboratory-made two circles goniometer with the copper $K_{\alpha 1}$ radiation ($\lambda = 1.5405$ Å) provided by a 5 kW RIGAKU RU-200 generator with a vertical linear focus in combination with a quartz monochromator [10]. Reciprocal space maps (RSMs) and symmetric θ-2θ scans were recorded around the (0004) Bragg reflection of GaN ($2\theta_B = 73.035°$). In this configuration, the penetration depth of X-rays in GaN is approximately 8 μm, which is significantly larger than the extension of the implanted area. In order to determine the strain along the direction normal to the surface, the θ-2θ curves are plotted versus $q[0001]/H(0004)$, where $q[0001]$ is the component of the deviation vector q along the normal [0001] to the surface, from the reciprocal lattice vector $H(0004)$ of the (0004) planes [11]. From the derivation of the Bragg's law, the normal strain can consequently be written as $\varepsilon_N = -q[0001]/H(0004)$.

RESULTS AND DISCUSSION

The damage evolution in the bulk and at the surface due to Eu implantation at RT seems to be intrinsic to GaN, since it has already been observed under different conditions such as 300 keV Au implantation at RT [3] or 1 MeV Au irradiation [12]. This behavior is still misunderstood and strongly differs from other semiconductors such as Si or AlN [13,14], where the accumulation of damage leads to an amorphization in the bulk.

Evolution of the strain state

The scattered X-ray intensity in the vicinity of the (0004) reflection of GaN is given by the RSMs recorded before implantation (Fig. 1(a)) and after Eu implantation at 5×10^{14} Eu/cm^2 (Fig. 1(b)). For comparison, the intensity is plotted in logarithmic scale using the same color scale. The intensity positions are defined by the normalized deviations from the reciprocal lattice vector along the normal direction to the surface [0001] and along [11$\bar{2}$0]. For both samples, a strong signal is observed at coordinates (0;0) which corresponds to the unperturbed region of the crystal. After implantation, an intense streak appears and the intensity spreads preferentially along the [0001] direction. This indicates a significant modification of the strain along the normal to the surface due to the formation of irradiation-induced defects in the implanted layer.

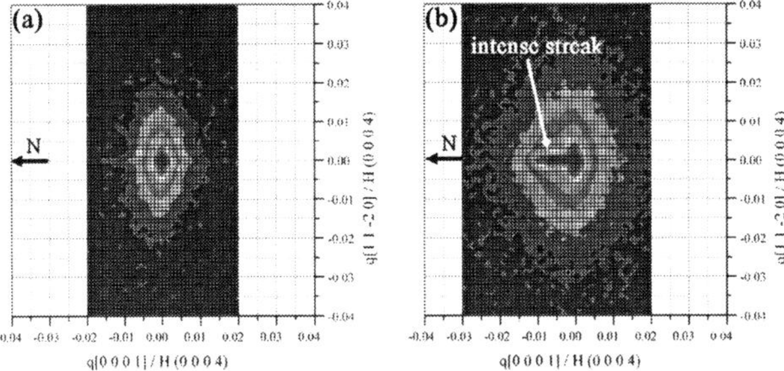

Figure 1. Reciprocal space maps close to the (0004) Bragg reflection of GaN (a) before implantation and (b) after 5×10^{14} Eu/cm^2 implantation (300 keV, RT). The intensity is plotted in logarithmic scale, in the same color scale. N indicates the surface normal direction.

Figure 2 shows the experimental θ-2θ scans acquired around the (0004) reflection of GaN before and after Eu implantation. The intensity is plotted versus $q[0001]/H(0004)$ in order to quantify the normal strain. For each sample, the intense peak at $q[0001]/H(0004)=0$ corresponds to the unperturbed part, and it is taken as an internal gauge for the strain determination.

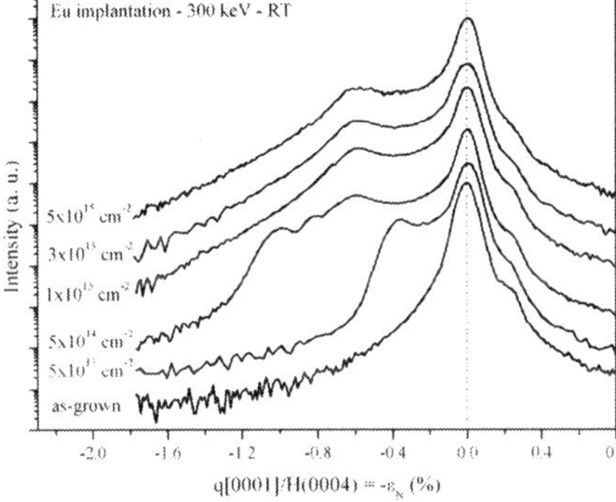

Figure 2. Evolution of the X-ray scattered intensity along the surface normal close to the (0004) reflection after Eu implantation (300 keV, RT) at various fluences.

Due to implantation damage, the intensity is scattered towards the negative $q[0001]/H(0004)$ values. At 5×10^{13} Eu/cm^2, a satellite peak is observed at $q[0001]/H(0004)\approx-0.4\%$ in addition to the peak of the undamaged region. This peak reveals a lattice expansion of the crystal along the surface normal. The presence of such a single and well-defined peak may indicate that the implanted region is quasi-homogeneously-strained with a mean strain level of 0.4%. At 5×10^{14} Eu/cm^2, a fringe-like pattern and a maximum normal strain of about 1.0% are observed. At higher fluences, from 1×10^{15} Eu/cm^2, the fringes pattern has disappeared. In the range of 1×10^{15} Eu/cm^2 to 5×10^{15} Eu/cm^2, it is worth to note that there is no significant evolution of the XRD curves: the maximum strain reaches approximately 1.6%, while a satellite peak at ε_N =0.6% is still present.

Evolution of the damage profile

The RBS/C spectra show a typical profile of damage after Eu implantation at room temperature, which is characterized by the presence of two peaks. One is related to the surface damage and the other one corresponds to the damage in the bulk. The relative defect concentration profiles, shown on Fig. 3, have been obtained from the RBS/C random and aligned spectra. Obviously, the level of damage increases with the fluence. Different regimes can be identified. At low fluence ($\leq 6\times10^{14}$ Eu/cm^2), the maximum of damage at the surface and in the bulk increases slightly but remains very low (≤ 0.1). Then, up to 1.2×10^{15} Eu/cm^2, the damage at the surface and in the bulk increases dramatically to reach 0.6. Above 1.2×10^{15} Eu/cm^2, the damage in the bulk saturates, while it continues to increase at the surface up to 0.9 and then saturates above approximately 3×10^{15} Eu/cm^2 [9].

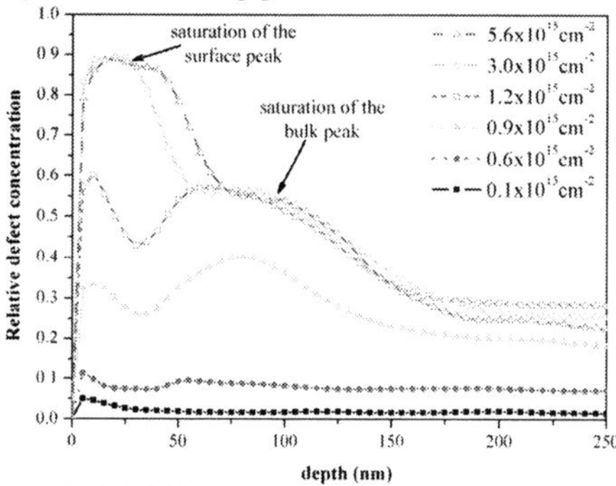

Figure 3. Damage profile from the RBS/C spectra of GaN after Eu implantation (300 keV, RT) at various fluences.

TEM observations

In order to understand the saturation observed by XRD and RBS/C around 1×10^{15} Eu/cm^2, TEM experiments have been carried out on two samples, before and after the saturation regime. Figure 4 summarizes the cross-section TEM images of the 5×10^{14} Eu/cm^2 sample acquired using different diffraction conditions. As shown in bright field conditions along the [$1\bar{1}20$] zone axis, a perturbed area of approximately 190 nm in width is observed from the surface (Fig. 4(a)). It is composed of three regions: one layer of dark contrast (B) of 80 nm in width, surrounded by two brighter layers of 40 nm at the surface (A) and 70 nm in the underlying part (C). Using the **g**=$1\bar{1}00$ weak beam conditions (Fig. 4(b)) in dark field, bright non-continuous lines parallel to the surface are visible (see black arrows). They reveal displacements in the basal planes and they correspond to basal stacking faults (BSFs), which are mainly located in the highly perturbed region (B). With the **g**=0002 weak beam conditions, another type of contrast in form of large bright dots is visible (Fig. 4(c)): it reveals the local atomic displacement along [0001] and it is related to clusters of point defects which are mainly located in the B region, but also in the A and C layers at lower density.

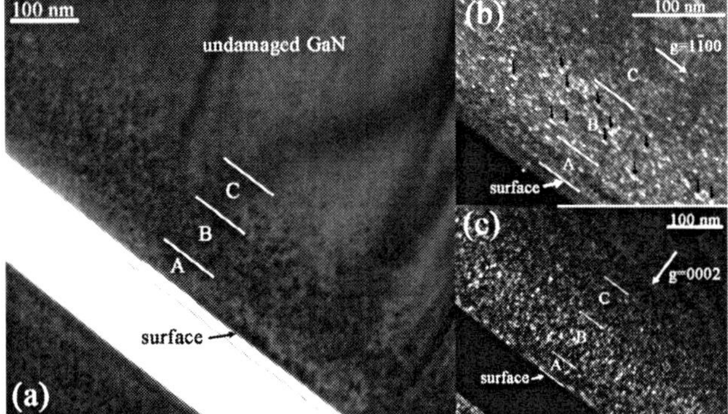

Figure 4. Cross-section TEM images of GaN implanted with 300 keV Eu ions at 5×10^{14} Eu/cm^2. (a) Bright field along the [11$\bar{2}$0] zone axis. Dark field using the (b) **g**=1̄100 and (c) **g**=0002 weak beam conditions.

The damaged area in the 3×10^{15} Eu/cm^2 sample extends now to the bulk up to a depth of 210 nm, as shown in the cross-section image seen along [11$\bar{2}$0] (Fig. 5(a) and Fig 5(b)).

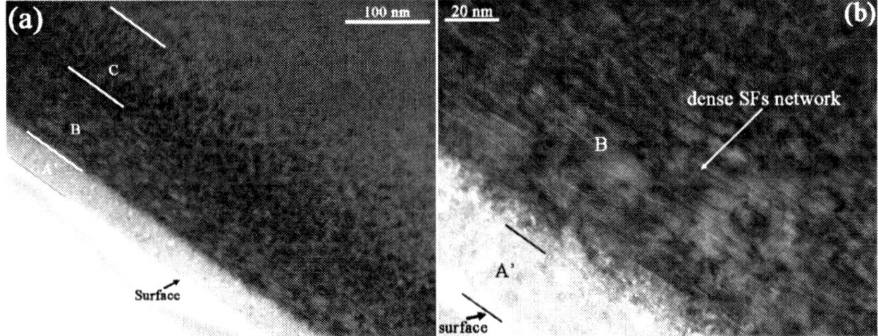

Figure 5. Cross-section TEM micrograph (along [11$\bar{2}$0]) of GaN implanted with 300 keV Eu ions at 3×10^{15} Eu/cm^2. (a) Three layers visible in the damaged area: (A') the nanocrystalline layer, (B) the dense network of SFs (also visible in (b) at higher magnification) and (C) the deep layer with defect clusters.

Three layers are clearly distinguished. A nanocrystalline layer of approximately 30 nm thick (A') is observed at the surface. It consists of small misoriented grains of less than 10 nm in size [4]. Then follows a layer revealing a dense SFs network (B), and the slightly less damaged underlying part (C) where the stacking faults are not dominant. Each of the latter two layers is about 90 nm thick. A detailed investigation has shown previously that the network of stacking faults is made of both basal and prismatic SFs [4]. The B and C layers are similar to those of the $5x10^{14}$ Eu/cm^2 sample. The upper part of the C layer (close to the B layer) exhibits a dark contrast, whereas the deepest part (close to the unperturbed crystal) is brighter and is composed of dark dots. Thus, the C layer most probably contains a non homogeneous distribution of point defect clusters.

SUMMARY

The above results show that the implantation damage formation in GaN exhibits at least three saturation stages versus the implantation fluence. Around $5x10^{14}$ Eu/cm^2, the strain state first levels off close to 0.6%. It is followed by a second saturation regime which occurs above a threshold of $1x10^{15}$ Eu/cm^2 as shown by both XRD and RBS/C. The damage profiles indicate that the defect saturation takes place in the bulk and it may be localized in the SFs region (layer B). There is no additional evolution of the strain state from $1x10^{15}$ Eu/cm^2. Interestingly, this second saturation regime takes place just before the formation of the nanocrystalline layer at the surface ($>2x10^{15}$ Eu/cm^2 [4]), and it seems to start when the SFs have reached the surface. Therefore, the first saturation may correspond to the initiation of the SFs propagation toward the surface. These two saturation regimes observed in GaN correspond to what has been known as an efficient dynamic annealing: migration, recombination and formation of defect complexes, which, in this material, allows the accommodation of larger implantation damage before the structural breakdown.

REFERENCES

1. A. J. Steckl, J. Heikenfeld, D. S. Lee and M. Garter, Mat. Sci. and Eng. B **81**, 97 (2001).
2. D. S. Lee and A. J. Steckl, Appl. Phys. Lett. **80**, 1888 (2002).
3. S. O. Kucheyev, J.S. Williams, C. Jagadish, J. Zou and G. Li, Phys. Rev. B **62**, 7510 (2000).
4. F. Gloux, T. Wojtowicz, P. Ruterana, K. Lorenz, and E. Alves, J. Appl. Phys. **100**, 073520 (2006).
5. V. Potin, P. Ruterana and G. Nouet, J. Phys. Condens. Matter **12**, 10301 (2000).
6. P. Ruterana, G. Nouet, Phys. Stat. Sol.(b) **227** , 177 (2001).
7. P. Ruterana, B. Lacroix and K. Lorenz, J. Appl. Phys. **109**, 013506 (2011).
8. J. F. Ziegler, J. P. Biersack, and U. Littmark, *The Stopping and Range of Ions in Solids*, (Pergamon, New York, 1985).
9. K. Lorenz, N. P. Barradas, E. Alves, I. S. Roqan, E. Nogales, R. W. Martin, K. P. O'Donnell, F. Gloux and P. Ruterana, J. Phys. D: Appl. Phys. **42** 165103 (2009).
10. S. Leclerc, A. Declémy, M. F. Beaufort, C. Tromas, and J. F. Barbot, J. Appl. Phys. **98**, 113506 (2005).
11. A. Debelle and A. Declémy, Nucl. Instrum. Meth. Phys. Res. B **268**, 1460 (2010).
12. W. Jiang, W. J. Weber, L. M. Wang and K. Sun, Nucl. Instrum. Meth. Phys. Res. B **218**, 427 (2004).
13. F. Gloux, P. Ruterana, K. Lorenz and E. Alves, Phys. Stat. Sol. (a) **205**, 68 (2008).
14. K. Lorenz, E. Alves, F. Gloux, P. Ruterana, M. Peres, A. J. Neves, and T. Monteiro, J. Appl. Phys. **107**, 023525 (2010).

Insulating Materials: Lasers and Phosphors

Mater. Res. Soc. Symp. Proc. Vol. 1342 © 2011 Materials Research Society
DOI: 10.1557/opl.2011.1072

Upconversion In Rare-Earth Ion-Doped NaYF₄ Crystals and Nanocolloids

Darayas N. Patel, Lauren A. Hardy, Tabatha J. Smith, Eva S. Smith and Donald M. Wright III
Oakwood University, Department of Mathematics & Computer Science,
7000 Adventist Blvd. Huntsville, AL 35896, USA.

ABSTRACT

Nano-colloids and nano-crystals doped with ions of rare-earth elements have recently attracted a lot of attention of scientific community. This attention is due to unique physical, chemical and optical properties attributed to nanometer size of the particles. They have great potential of being used in applications spanning from new types of lasers, especially blue and UV lasers, phosphorous display monitors, optical communications, and fluorescence imaging. In this paper we investigate the infrared-to-visible upconversion luminescence in bulk crystals doped with ytterbium and holmium co-doped and ytterbium and thulium co-doped NaYF₄ upconversion phosphors. The phosphors were prepared by using simple co-precipitation synthetic method. The initially prepared phosphor has very weak upconversion fluorescence. The fluorescence significantly increased after the phosphor was annealed at a temperature of 600 ^{0}C. Nanocolloids of this phosphor were obtained using methanol as solvents and they were utilized as laser filling medium in photonic crystal fibers. Under 980 nm laser excitation very strong upconversion signals were obtained for ytterbium and holmium co-doped phosphor at 541 nm, 646 nm and 751 nm, and 376 nm, 476 nm, 646 nm, 696 nm and 803 nm for ytterbium and thulium co-doped phosphor. The particle sizes of the nanocolloids were analyzed using Atomic Force Microscope. The reported nanocolloids are good candidates for fluorescent biosensing applications and also as a new laser filling medium in fiber lasers.

INTRODUCTION

Nano-colloids and nano-crystals doped with ions of rare-earth elements have recently attracted a lot of attention from the scientific community. This attention is due to unique physical, chemical and optical properties attributed to nanometer size of the particles. They have great potential of being used in applications spanning from new types of lasers, especially blue and UV ones, phosphorous display monitors, optical communications, fluorescence imaging and novel approaches in vivo imaging, disease detection and diagnosis[1-4]. Compared to the conventional fluorophores like dyes, upconverting phosphors have many advantages as fluorescence labeling materials, such as low photobleaching, absence of autofluorescence, low back ground noise, high sensitivity for detection and feasibility of multiple labeling with different emissions under the same laser excitation. Upconversion is a process where low-energy light, usually near-infrared or infrared is converted to high energy light (UV or visible), through multiple photon absorptions or energy transfers. Upconversion emission of rare-earth ion doped nanocrystals has been observed in several different host materials. Up to now, hexagonal phase NaYF₄ has been regarded as the most efficient upconversion host material.

Trivalent holmium ion is of great interest due to its strong green upconversion luminescence and the presence of trivalent ytterbium ions as an ideal co-dopant are well known to increase the optical pump efficiency of the luminescence in materials because the Yb^{3+} have a large absorption cross section around 980 nm and can efficiently transfer the excitation energy to other rare earth ions. The fluoride hosts studied in our current work are strong and efficient upconverters because of their low phonon energy as compared to the phosphates and oxides thus increasing the number and the probability of radiative transitions in rare-earth doped materials.

In this work, we report the upconversion luminescence properties of Ho^{3+}/Yb^{3+} co-doped sodium yttrium fluoride and Tm^{3+}/Yb^{3+} co-doped sodium yttrium fluoride. The particle sizes of the nanocolloids were analyzed using Atomic Force Microscope. These most efficient hexagonal phase co-doped $NaYF_4$ visible upconversion phosphor materials will be extensively utilized as possible candidates for laser filling medium in micro-structured fiber lasers.

EXPERIMENTAL

<u>Synthesis of the Sodium Yttrium Fluoride Crystals Co-Doped with Rare-Earth</u>

Many rare earth doped materials of different composition, shapes and sizes distribution have been prepared by different kinds of synthetic methods such chemical deposition, sol-gel process, micro-emulsion techniques, gas phase condensation methods, hydrothermal methods and laser ablation[6]. All of those methods have been investigated for the production of luminescent nanomaterials. However, in this particular experiment, the method known as solution based co-precipitation techniques was utilized. The $NaYF_4$: Ho^{3+}, Yb^{3+} crystals were prepared in the presence of Na_2-ethylenediaminetetraacetic acid (EDTA) using the co-precipitation procedure. The solution was prepared by dissolving 0.2mol of ytterbium chloride in 4mL of water, 0.2 mol of $HoCl_3$ in 1mL of water, 0.5 mol of NaF in 60mL of water, 0.2 mol of EDTA in 20mL of water. The YCl_3 solution was obtained by dissolving Y_2O_3 in hydrochloric acid and adjusting to pH 2 to avoid any hydrolysis. The rare-earth chloride solution was allowed to mix with the EDTA solution for metal-EDTA complex to occur. The EDTA complex solution was quickly introduced into the NaF solution and the mixture was allowed to stir vigorously for several hours. After stirring the solution was allowed to sit overnight for the powder particles to settle. Powder particles from the reaction were centrifuged, washed three times with distilled water and then once with ethanol. The particles were dried under vacuum to remove any traces of water. The dried white powder weighed about 1.0 g. The $NaYF_4$: Yb^{3+}, Ho^{3+} crystals prepared in the above procedure did not show any upconversion fluorescence. However, after the dried precipitate was annealed to a temperature of 600°C for a period of one hour bright green upconversion was observed under a 980nm diode laser excitation. The $NaYF_4$: Tm^{3+}, Yb^{3+} crystals were also prepared using the co-precipitation procedure. The 0.2 mol of $HoCl_3$ was replaced with 0.2 mol $TmCl_3$. Tm^{3+} ions have been selected because it has some metastable levels suitable for emitting blue and UV luminescence.

<u>Preparation of the Nanocolloids of Sodium Yttrium Fluoride</u>

Several methods have been used to prepare rare-earth nanocolloids, such as liquid-phase synthesis in high boiling point coordinating solvents[8] followed by dispersing the synthesized nanocrystals into polar and non-polar solvents[9], wet chemical synthesis[1,3] and two step dispersion process[5]. Several papers have been reported on upconversion luminescence of transparent rare-earth-ion-doped halide nanocolloids[2]. In this study, synthesized $NaYF_4$ crystals co-doped with trivalent ions of holmium and ytterbium was used for making nanocolloids in methanol. In preparation of the nanocolloids, a small amount of the synthesized $NaYF_4$ crystals was added to 15-20 ml of methanol and stirred vigorously for several hours. The samples were left undisturbed for several days for the large size particles to settle at the bottom. Clear liquids over precipitate were collected with a syringe and using a micron disk filter to obtain the nanocolloid.

Atomic Force Microscope Analysis

The particle sizes of the nanocolloid were analyzed using the Pico Plus Atomic Force Microscope (AFM) from Molecular Imaging (now Agilent Technologies). The AFM creates a three dimensional representation of the sample surface by monitoring the force of interaction between the sample and the probe (cantilever). AFM can obtain topographical images of nanomaterials in their natural state without surface treatment or coating. The $NaYF_4$: Yb^{3+}, Ho^{3+} nanocolloids were characterized in AFM taping mode. A few drops of nanocolloid were placed on freshly pealed mica substrate and allowed to dry. 3D topographically images of the $NaYF_4$:Yb^{3+}, Ho^{3+} nanoparticles are presented in Fig. 1. This AFM analysis indicated that the average size (average mean height) of the nanoparticles in the colloid was close to 274.2 nm.

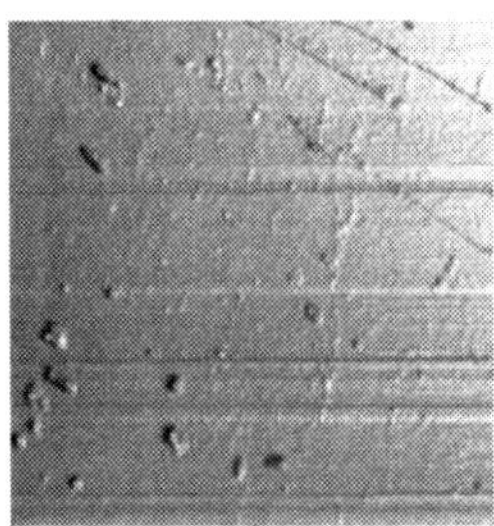

Fig. 1 Topographically AFM image of the NaYF4: Yb 3+, Ho3+ nanoparticles

Emission Spectroscopy

Upconversion luminescence in bulk crystals was conducted using a 980nm diode laser from Coherent as pumping sources. In all the measurements, the sample was at room temperature. Optical fluorescent spectra were taken with the Princeton Instruments 500nm focal length Spectra Pro (Sp-2500i) imaging spectrometer/monochromator equipped with 1200 gr/mm blazed at 500 nm holographic diffraction grating, R929 photomultiplier tube, and PI-Max 102HQ Digital Intensified CCD Camera system. The crystalline powder was pressed into flat disks to obtain the emission spectra. The emission measurements were made in reflectance mode using a sample chamber with the sample placed approximately 45° off the entrance slit focal axis. Under 980 nm laser excitation very strong upconversion signals were obtained for ytterbium and holmium co-doped phosphor at 541 nm (green) ($^5S_2 \rightarrow {}^5I_8$), 646 nm (red) ($^5F_5 \rightarrow {}^5I_8$) and 751 nm (red) ($^5I_4 \rightarrow {}^5I_8$), and 376 nm (UV)($^1D_2 \rightarrow {}^3H_6$), 476 nm (blue)($^1G_4 \rightarrow {}^3H_6$), 646 nm (red) ($^3F_2 \rightarrow {}^3G_6$), 696 nm (red) ($^3F_3 \rightarrow {}^3H_6$) and 803 nm (red) ($^3F_4 \rightarrow {}^3H_6$) for ytterbium and thulium co-doped phosphor.

RESULTS & DISCUSSION

The upconversion excitation pathways for the Ho^{3+}/Yb^{3+} and Tm^{3+}/Yb^{3+} ions couples in these materials are well known[9] and are shown in energy level diagram in figure 2. In the case of the $NaYF_4$: Ho^{3+}, Yb^{3+}, an initial energy transfer from the Yb^{3+} ion in the $^2F_{5/2}$ state to an Ho^{3+} populates the $^5I_5$ level. A second 980 nm photon, or energy transfer from the Yb^{3+} ion, can populate the $^5S_2$ of the Ho^{3+} ion. The green $^5S_2 \rightarrow ^5I_8$ emission occurs. The ion relaxes nonradiatively from $^5S_2$ to the $^5F_5$ level and red $^5F_5 \rightarrow ^5I_8$ emission occurs and the ion relaxes nonradiatively from $^5F_5$ level to the $^5I_4$ level and another red $^5I_4 \rightarrow ^5I_8$ emission occurs. Fig. 3 shows the emission spectra of the Ho^{3+}, Yb^{3+}:$NaYF_4$ under 980 nm laser excitation. In order to determine the number of photons responsible for the upconversion mechanism, the intensity of the upconversion emission was recorded as a function of the 980 nm diode laser excitation intensity. The 541 nm, 646 nm and 751 nm upconversion signals demonstrated a near-quadratic pump power dependency indicating that two photons were involved in generating the 541 nm, 646 nm and 751 nm upconversion emissions. Fig. 4 shows quadratic power dependence of the 541 nm upconversion signal with 980 nm diode laser excitation. For the Tm^{3+}, Yb^{3+}:$NaYF_4$ sample, up to four subsequent energy transfers from Yb^{3+} ions populate the upper Tm^{3+} levels and several emissions can occur. Fig. 5 and 6 shows the emission spectra of the Tm^{3+}, Yb^{3+}:$NaYF_4$ under 980 nm laser excitation.

Optical fluorescent spectroscopy of the nano-colloids was conducted using an air cooled Argon ion laser. The nano-colloid samples were placed in 3.5 ml glass fluorometric cuvettes mounted inside sample chamber of the monochromator. When the Ho^{3+}, Yb^{3+}:$NaYF_4$ colloid was excited with an argon ion laser emission lines were obtained at 369 nm, 549 nm, 708 nm, 757 nm and 866 nm.

Recently, there have been attempts of building laser medium by filling micro-structured optic fibers with solutions of fluorescent materials[10]. Our team is planning to test this concept by filling the micro-structured fibers with nano-colloids of Ho^{3+}, Yb^{3+}:$NaYF_4$. The nano-colloid will be driven in the holes of the fiber by capillary forces. The nano-particles of $NaYF_4$ are evenly dispersed along the inner walls of the holes of the fiber. This will make up an optically pumped laser/amplifier. Radiation from an external pumping laser will be concentrated in the holes of the fiber and excite the ions of rare-earth in the nano-particles thus potentially producing laser effect in the fiber.

One of the major enhancement factors was annealing the phosphor at a temperature of 600^0C which significantly increased the fluorescence. Another important enhancement factor was excitation of these phosphor materials with inexpensive commercial 980 nm laser diodes.

CONCLUSIONS

Ho^{3+}, Yb^{3+}:$NaYF_4$ the most efficient known infrared-to-visible upconversion phosphor, was synthesized using the simple co-precipitation procedure in the presence of EDTA. After annealing at a temperature of 600^oC very strong green upconversion fluorescence was visible to the naked eye. The Ho^{3+}, Yb^{3+} co-doped crystals and the Tm^{3+}, Yb^{3+} co-doped crystals were capable of upconverting NIR light from a 980 nm diode laser into green and red fluorescence and UV, blue and red respectively.

The synthesis does not require any sophisticated equipment or complicated procedures. The synthesized upconversion phosphor can be excited with inexpensive commercial 980 nm laser diodes. This dramatically increases its commercialization potential. The upconverting phosphors are good candidates for biolabels in biological assays. Nanocolloids of $NaYF_4$ have been obtained and their possible incorporation into the micro-structured optical fiber media for fiber laser applications.

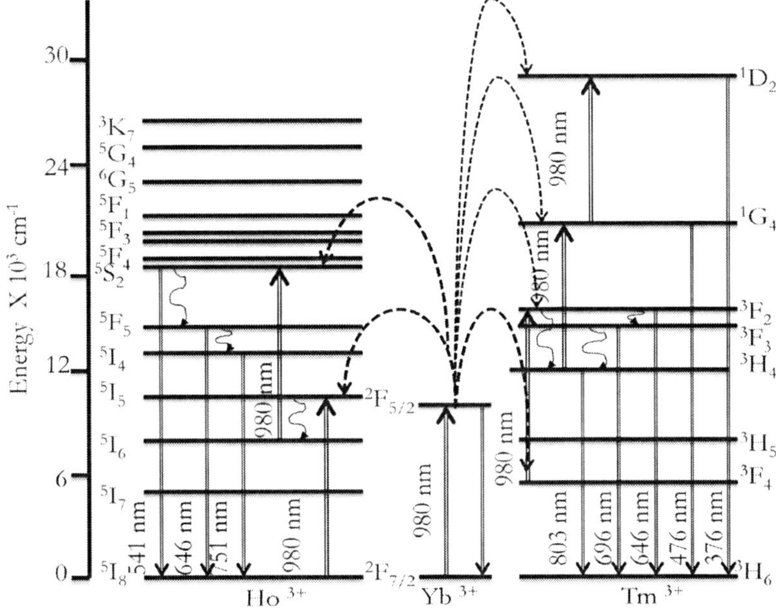

Fig. 2 Partial energy level diagram of NaYF₄

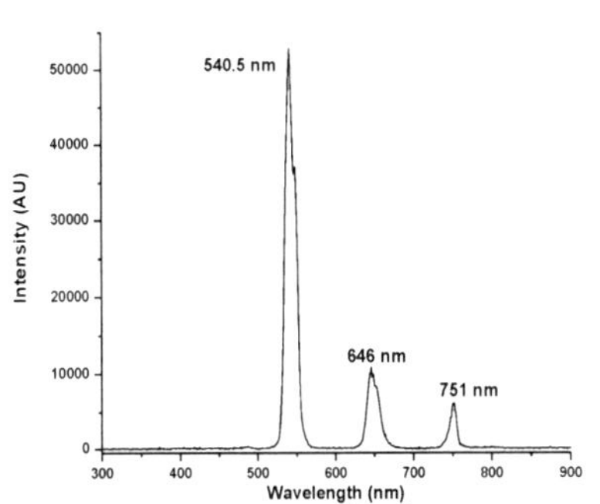

Fig. 3 Emission spectra of NaYF₄: Ho³⁺ Yb³⁺ with EDTA baked at 600⁰C excited with a 980 nm diode laser

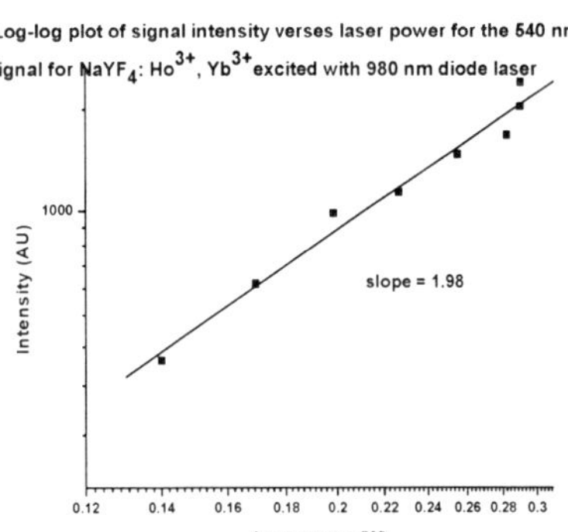

Fig. 4 Power dependence of 540 nm upconversion signal under 980 nm laser excitation

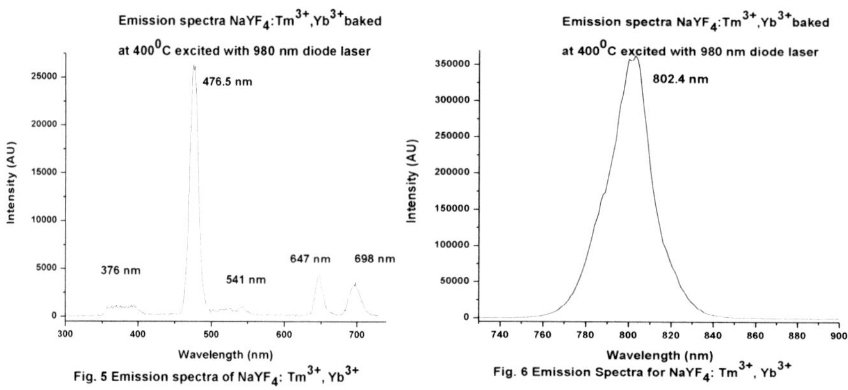

Fig. 5 Emission spectra of NaYF$_4$: Tm^{3+}, Yb^{3+}

Fig. 6 Emission Spectra for NaYF$_4$: Tm^{3+}, Yb^{3+}

The Ho^{3+}, Yb^{3+}:NaYF$_4$ nanocolloids were showed to be good candidates for fluorescent application and also as a new laser filling medium in fiber lasers[11].

ACKNOWLEDGMENTS

This work was supported by the STEMER grant from National Science Foundation.

REFERENCES

1. R. Scheps, Prog. Quant. Electron. 20(4), 271-358, (1996).
2. J. S. Chivian, W. E. Case, and D. D. Edden, Appl. Phys. Lett. 35(2), 125, (1979).
3. E. Downing, L. Hesselink, J. Raltson and R. Macfarlane, Science 273, 1185, (1996).
4. R. S. Niedbala, H. Feindt, K. Kardos, T. Vail, J. Burton, B. Bielska, S. Li, D. Milunic, P. Borton, P. Bourdelle and R. Vallejo, Anal. Biochem. 293(1), 22, (2001).
5. J. Suyer, A. Aebischer, D. Biner, P. Gerner, J. Grimm, S. Heer, K. W. Kramer, C. Reinhard and H. U. Gudel, Optical Materials **27**, 1111-1130 (2005).
6. G. Yi, H. Lu, S. Zhao, Y. Ge, W. Yang, D. Chen and L. Guo, Nano Letters, Vol. **4**, No. 11, 2191-2196 (2004).
7. K. Riwotzki, H. Meyssamy, A. Kornowski and M. Haase, J. Phys. Chem. B. **104**, 2824-2828 (2000).
8. Y. Wang, W. Qin, J. Zhang, C. Cao, J. Zhang and Y. Jin, J. Rare-Earths **26**, 40-43 (2008).
9. J. Boyer, L. A. Cuccia and J. A. Capobianco, Nano Lett., Vol. **7**, No. 3, , 847-852, (2007).
10. C. Kerbage, R. S. Windeler, B. J. Eggleton, P. Mach, M. Dolinski and J. A. Rogers, "Electrically driven motion of micro-fluids in air-silica microstructured fiber: application to tunable filter/attenuator," Optical Fiber Communication Conference (OFC) Anaheim, California, Microstructured Fibers and Photonic Crystal Device (ThK) March 17, (2002).
11. D. Patel, C. Vance, N. King, M. Jessup, L. Green and S. Sarkisov, Proceedings of SPIE Vol. 75598, 759821 (**2010**).

Mater. Res. Soc. Symp. Proc. Vol. 1342 © 2011 Materials Research Society
DOI: 10.1557/opl.2011.781

Preparation of luminescent inorganic core/shell-structured nanoparticles

Moritz Milde[1], Sofia Dembski[1], Sabine Rupp[1], Carsten Gellermann[1], Gerhard Sextl[1,2], Miroslaw Batentschuk[3], Andres Osvet[3], Albrecht Winnacker[3]

[1] Fraunhofer Institute for Silicate Research, Neunerplatz 2, 97082 Wuerzburg, Germany
[2] Department of Chemical Technology of Materials Synthesis, University of Wuerzburg, Roentgenring 11, 97070 Wuerzburg
[3] Chair WW6 Materials for Electronics and Energy Technology (i-MEET), University of Erlangen-Nuremberg, Martensstr. 7, 91058 Erlangen, Germany

ABSTRACT

SiO_2 nanoparticles (NPs) were coated with Eu^{3+}-doped calcium phosphate (CP) and Mn^{2+}-doped ZnO to give Zn_2SiO_4 via a modified Pechini sol-gel process. Annealing at high temperatures resulted in NPs with an amorphous core and a crystalline luminescent shell. It was shown that this procedure can be applied to silica cores with diameters below 300 nm. By transmission electron microscopy, elemental analysis and from X-ray diffraction patterns it was determined that shell composition and structure are influenced by the annealing temperature and pH of the coating solution. Measurements of photoluminescence intensities displayed their dependency on the concentration of dopant in the resulting core/shell NPs.

INTRODUCTION

Luminescent inorganic NPs on the basis of non-semiconductor materials have gained increasing interest over the past few years from scientists mainly in fields of applied research. Areas of potential applications include optics and optoelectronics or biological and medical diagnostics [1,2]. Because of excellent optical characteristics such as long luminescence lifetimes, narrow emission bands and a high photostability, these NPs promise to be an alternative to toxic semiconductor NPs (quantum dots, QDs) for biological labeling applications [3].

The preparation of dispersible NPs with a defined structure, size and surface texture still remains a challenge. During the last decade, several wet-chemical methods have been developed to produce isolated luminescent inorganic NPs e.g. sol-gel-synthesis, solvothermal reaction, co-precipitation and microemulsion methods [4,5]. Each of these methods has its own applicability, advantages and disadvantages. Here, we present the preparation of core/shell-structured inorganic luminescent NPs utilizing a Pechini-type sol-gel process. This process was already applied to a variety of phosphors e.g. Zn_2SiO_4, $LaPO_4$, YVO_4 [6,7]. However, up to now sizes of the silica cores used were > 300 nm. It was shown that the formation of an inorganic shell is also possible for cores below that size. Moreover this approach was adapted to other systems such as calcium phosphate (CP). Using this strategy it is possible to obtain spherical NPs with a narrow size distribution predefined by the silica cores. Only a thin layer of phosphor material is needed to determine the optical properties of these NPs. Due to their chemical and physical properties the core/shell systems $SiO_2/Ca_{10}(PO_4)_6(OH)_2$ and SiO_2/Zn_2SiO_4 are promising candidates for deployment as luminescent markers in biological and medical diagnostics.

Characterization of NPs was done by X-ray diffraction analysis (XRD), Fourier transform infrared spectroscopy (FT-IR), transmission electron microscopy (TEM), dynamic light scattering (DLS) and inductively coupled plasma optical emission spectrometry (ICP-OES). Luminescent properties were investigated by photoluminescence spectroscopy (PL).

EXPERIMENTAL PROCEDURE

SiO_2 cores were synthesized according to a modified Stoeber process [8]. $SiO_2/CP:Eu^{3+}$ core/shell, $SiO_2/ZnO:Mn^{2+}$ and $SiO_2/Zn_2SiO_4:Mn^{2+}$ NPs were synthesized as described previously [9,10]. Precursor salts of shell components were dissolved in a mixture of ethanol and water. Citric acid was used to chelate the metal ions and polyethylene glycol (PEG) as a crosslinking agent. The pH of solutions was adjusted by using nitric acid or ammonium hydroxide. The silica cores dispersed in ethanol were added to the coating solution and the mixtures were stirred at room temperature. The coated NPs were sedimented by centrifugation, redispersed in water, freeze-dried and annealed in the temperature region from 800 to 1100 °C.

The morphology of the core/shell NPs was studied with a Zeiss EM 10 with an acceleration voltage of 80 kV. Crystallinity of the powder samples was analyzed using a Philips PW 1152. ICP-OES was done by a Varian Wista Pro. Emission and excitation spectra of $SiO_2/ZnO:Mn^{2+}$ and $SiO_2/Zn_2SiO_4:Mn^{2+}$ were recorded using a xenon lamp (70 W) with an interference filter; the luminescence was dispersed with a Jobin-Yvon HRS-2 monochromator with a resolution of 0.5 nm and detected with an R-453 photomultiplier.

DISCUSSION

SiO_2 NPs were coated with an inorganic luminescent shell according to a modified Pechini sol-gel procedure as outlined in figure 1. Size and morphology of the NPs are predefined by silica cores. Adsorption of metal ions and the anionic counterpart on the silica surface is accomplished through silanol-groups. During the coating procedure citric acid is utilized to form complexes with the metal ions. Additionally an organic framework is built up through polymerization of citric acid and PEG [7,11]. In this way a homogeneous distribution of shell components is achieved because the metal complexes are immobilized and segregation of particular metal ions is reduced. By subsequent annealing the organic network is decomposed and at elevated temperatures a crystalline luminescent shell is formed around the silica cores [7].

In the course of our investigations it was found that the variation of synthesis conditions during the coating process can significantly change structural and luminescent properties of the resulting core/shell NPs. Three important parameters that affect the synthesis of $SiO_2/Ca_{10}(PO_4)_6(OH)_2$, SiO_2/ZnO and SiO_2/Zn_2SiO_4 NPs were investigated in detail. For the surface coating of silica cores with a CP shell the pH during the coating step played a major part in shell formation. In the synthesis of SiO_2/Zn_2SiO_4 NPs it was determined that annealing temperature influences the ratio of ZnO to Zn_2SiO_4 in the shell. Optical properties were highly dependent on the doping concentration. This effect was analyzed for $SiO_2/ZnO:Mn^{2+}$ and $SiO_2/Zn_2SiO_4:Mn^{2+}$ core/shell NPs.

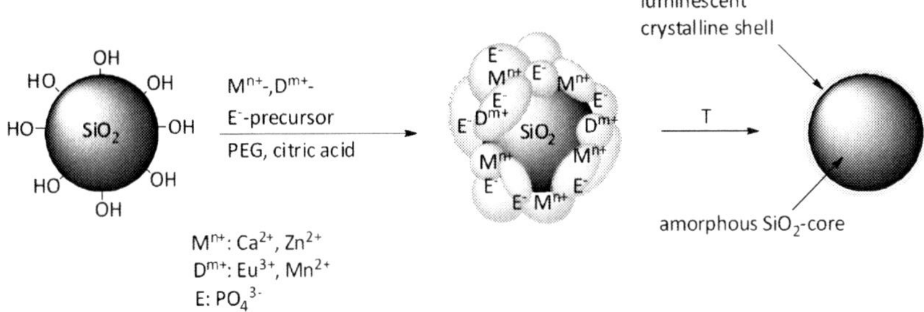

Figure 1. General approach to the preparation of core/shell-structured luminescent NPs.

Relationship between pH in reaction solution and calcium phosphate phase

Coating of SiO$_2$ NPs with a Eu^{3+}-doped CP-shell may be a possibility to obtain biocompatible markers with excellent optical properties. However, it was found out that in the synthesis of SiO$_2$/CP:Eu^{3+} the pH-value of the coating solution plays a major role for a successful coating. The difference in samples prepared at pH values of 3, 6 and 9 was investigated by TEM, XRD (figure 2) and ICP-OES.

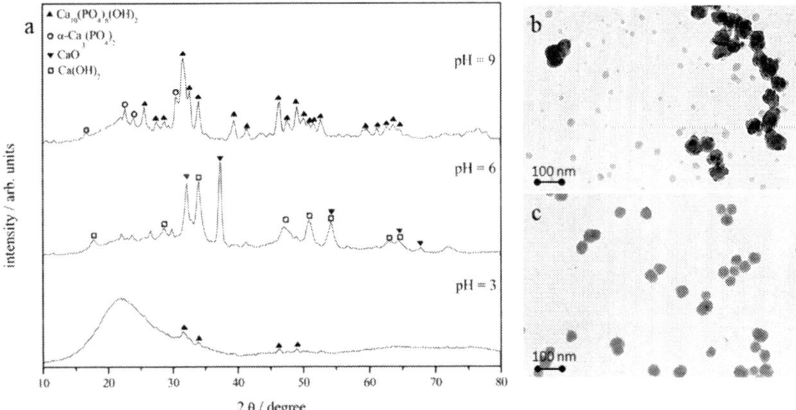

Figure 2. Left: XRD patterns of SiO$_2$/CP:Eu^{3+} core/shell NPs prepared at pH values of 3, 6 and 9. The phase composition was checked by means of JCPDS reference patterns for Ca$_{10}$(PO$_4$)$_6$(OH)$_2$ (PDF Ref. 74-0565), α-Ca$_3$(PO$_4$)$_2$ (PDF Ref. 29-0359), CaO (PDF Ref. 48-1467) and Ca(OH)$_2$ (PDF Ref. 04-0733).
Right: TEM images of SiO$_2$/CP:Eu^{3+} core/shell NPs prepared at pH 6 (b) and 9 (c).

At pH 3 only a very thin layer of hydroxyapatite ($Ca_{10}(PO_4)_6(OH)_2$, HAp) with poor crystallinity resulted on the silica cores. Carrying out the reaction at pH 6 led to the formation of calcium hydroxide $Ca(OH)_2$ and calcium oxide CaO. As can be seen in TEM image (figure 2b) secondary spherical NPs with diameters ranging from 2 to about 40 nm were formed and no coating of silica cores was achieved. Further increase of the pH to 9 resulted in the formation of a shell consisting of HAp and α-tricalcium phosphate (α-$Ca_3(PO_4)_2$, α-TCP). The amounts of calcium and europium detected in this sample were 52 % and 84 % relative to the precursor compounds. This indicated that the starting materials were not completely deposited on the silica surface. The ratio of calcium to phosphorus was found to be 1.60 which is consistent with a phase composition of HAp (1.67) and α-TCP (1.50) [12].

Relationship between annealing temperature and composition of zinc containing shell compounds

Mn^{2+}-doped α-Zn_2SiO_4 is a phosphor with high luminescent intensities and long lifetimes. In the synthesis of $SiO_2/Zn_2SiO_4{:}Mn^{2+}$ core/shell NPs the annealing temperature has a significant influence on the crystal phase in the shell. After exposing the freeze-dried NPs to 800 °C only zinc oxide could be identified by XRD analysis (figure 3). When increasing the temperature up to 900 or 1000 °C a mixture of willemite (α-Zn_2SiO_4) and ZnO was received. At a temperature of 1100 °C pure zinc silicate resulted.

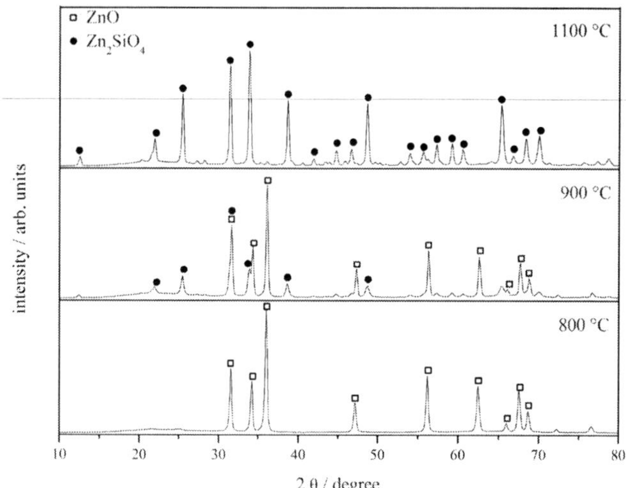

Figure 3. XRD patterns of $SiO_2/Zn_2SiO_4{:}Mn^{2+}$ core/shell NPs annealed at temperatures of 800 °C, 900 °C and 1100 °C. The phase composition was checked by means of JCPDS reference patterns for α-Zn_2SiO_4 (PDF Ref. 37-1485) and ZnO (PDF Ref. 36-1451).

These observations indicated that the formation of zinc silicate proceeds in two consecutive steps. First ZnO evolves from the zinc precursor and builds a coating around the silica cores. Ensuing diffusion processes from ZnO with incorporated Mn^{2+} into the silicate matrix lead to the development of Mn^{2+} doped Zn$_2$SiO$_4$. Photoluminescence measurements showed that the increase of the annealing temperature also led to a higher luminescence intensity.

Influence of doping concentration of Mn^{2+} on photoluminescent properties

An additional parameter that was varied in the synthesis of SiO$_2$/ZnO and SiO$_2$/Zn$_2$SiO$_4$ is the initial concentration of Mn^{2+} in the coating solution. Experiments showed that by increasing the doping concentration of Mn^{2+}, the intensity of emission with its maximum at 525 nm [13] could be raised notably (figure 4). The NPs were annealed at 900 °C so the shell in these samples exhibited a mixture of ZnO and α-Zn$_2$SiO$_4$. Target doping concentrations were 1 and 20 mol % Mn^{2+} relative to the amount of zinc, the effective amounts of Mn^{2+} incorporated into the samples accounted for 0.1 and 3.3 mol %. The latter showed a higher PL intensity which could be ascribed to the additional number of emitting Mn^{2+}.

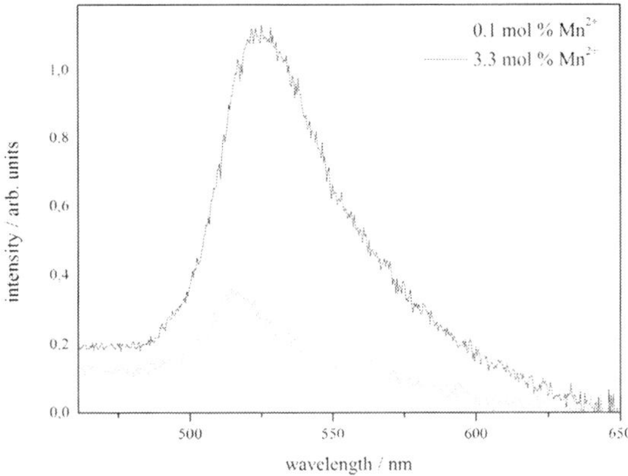

Figure 4. PL spectra of SiO$_2$/Zn$_2$SiO$_4$:Mn^{2+} core/shell NP samples annealed at 900 °C and doped with 0.1 mol % and 3.3 mol % Mn^{2+}.

The maximum of emission in the higher doped sample was slightly red shifted. The reason for this may be the distortions of the host lattice around Mn^{2+} or the formation of Mn-Mn-pairs due to the high doping concentration. The PL decay curve of the sample with 0.1 mol % recorded at 525 nm doping showed exponential behavior which is characteristic for the forbidden Mn^{2+} d-d-transition in willemite. Interestingly the corresponding curve of the higher doped sample exhibited a non-exponential decay which gave evidence for concentration quenching [14].

CONCLUSIONS

It was shown that the synthesis of core/shell-structured inorganic luminescent nanoparticles by a Pechini-type sol-gel process is possible for SiO_2 cores with sizes below 300 nm. The formation of the shell can be adjusted by variation of several reaction parameters: it is mainly the pH of the coating solution, the annealing temperature and the dopant concentration that have an effect on the composition, structural and optical properties of the particles. Furthermore, by carefully selecting the synthesis conditions the formation of a shell on the silica surface can presumably be adapted to other compositions.

ACKNOWLEDGMENTS

This work was financially supported by the Fraunhofer-Gesellschaft zur Foerderung der angewandten Forschung e.V., Munich, Germany (Grant No. 692034 and 663662) and the Deutsche Forschungsgemeinschaft DFG (Grant No. GE 2118/3-1 and BA 2245/3-1). Authors also gratefully acknowledge Prof. Dr. G. Krohne (Biocenter, University of Wuerzburg, Germany) for the use of the electron microscopes.

REFERENCES

1. H. A. Höppe, *Angew. Chem. Int. Ed.* **48**, 3572 (2009).
2. J. Shen, L.-D. Sun and C.-H. Yan, *Dalton Trans.* **42**, 5687 (2008).
3. S. P. Mondéjar, A. Kovtun and M. Epple, *J. Mater. Chem.* **17**, 4153 (2007).
4. H. Chander, *Mater. Sci. Eng.* **R 49**, 113 (2005).
5. R. Dittmeyer, R.W. Keim, G. Reysa and A. Oberholz, *Chemische Technik: Prozesse und Produkte. Band 2: Neue Technologien.* (Wiley-VCH, Weinheim 2004).
6. Y. Wu, Y. Wang, D. He, M. Fu, Z. Che and Y. Li, *J. Lumin.* **130**, 1768, (2010).
7. J. Lin, M. Yu, C. Lin and X. Liu, *J. Phys. Chem. C*, **111**, 5835 (2007).
8. W. Stöber, A. Fink and E. Bohn, *J. Colloid Interface Sci.* **26**, 62 (1968).
9. S. Dembski, S. Rupp, C. Gellermann, M. Batentschuk, A. Osvet and A. Winnacker, *J. Colloid Interface Sci.* (2010) in press.
10. S. Dembski, M. Milde, M. Dyrba, S. Schweizer and C. Gellermann, (2011) in progress.
11. M. Kakihana and M. Yoshimura, *Bull. Chem. Soc. Jpn.* **72**, 1427 (1999).
12. J. Pena and M. Vallet-Regi, *J. Eur. Ceram. Soc.* **23**, 1687 (2003).
13. A. L. N. Stevels and A. T. Vink, *J. Lumin.* **8** 443 (1974).
14. A. Morell and N. El Khiati, *J. Electrochem. Soc.* **140**, 2019 (1993).

Mater. Res. Soc. Symp. Proc. Vol. 1342 © 2011 Materials Research Society
DOI: 10.1557/opl.2011.1155

Vacuum Deposited Erbium-Doped NIR Luminescent Organic Thin Films For 1.5 µm Optical Amplification Applications

Laurent Divay[1], Christophe Galindo[1], Evelyne Chastaing[1], Renato Bisaro[1], Frédéric Wyczisk[1], Pierre Le Barny[1]

[1] THALES R&T, 1 avenue Augustin Fresnel, 91767 Palaiseau Cedex, France

ABSTRACT

Inorganic erbium-doped glasses are widely used in telecommunications due to the sharp intra-atomic $^4I_{13/2} \rightarrow {}^4I_{15/2}$ transition in the 4f orbital of erbium resulting in an emission at ~ 1.5 µm, which is the low loss window of silica optical fibres. The limited erbium concentration of about 10^{20} ions.cm^{-3} in inorganic erbium-doped glasses and the low absorption coefficient of the Er^{3+} ions, imply that relatively long lengths of fibre are required. The organic erbium complexes present higher absorption cross sections due to the photosensitization of erbium by organic conjugated ligands and broader emission bands than those of the free Er^{3+} ion. Such properties open the possibility to develop compact, low power and broadband infrared emitting devices. We present the study of a highly doped organic thin film obtained from organic erbium complexes deposited by a vacuum sublimation technique. This deposition method allows the realization of an erbium-doped thin film without the help of an organic polymer matrix, which is a potential source of vibrationnal luminescence quenching. The ligands used in the present study are fluorinated in order to limit the vibrationnal quenching of the ligand itself, and to increase the volatility of the complexes. In this paper, we report the synthesis, the sublimation process and the characterization of the thin films.

INTRODUCTION

Lanthanide materials attract considerable interest in the field of light emitting materials due to the sharp luminescence peak of most lanthanide ions. Owing to their unique photophysical properties, lanthanides are used in solid state NIR laser applications as dopant in inorganic crystals and glasses. Erbium doped glass fibres are extensively used in the telecommunication field due to the $^4I_{13/2}$-$^4I_{15/2}$ NIR transition of the Er^{3+} ion around 1.5 µm, in a low-loss window of silica optical fibres. Due to the limited solubility of the erbium ions in glasses, doping concentration is usually fairly low (less than 10^{20} atoms.cm^{-3}). In addition to the very small absorption and emission cross sections of the lanthanides, large interaction lengths are needed for optical amplification to take place. For this reason, erbium doped amplifiers are usually doped fibres that offer the interaction length needed for efficient amplification. However, planar optical amplifiers would find many applications in photonic integrated circuits. The challenge is then to fabricate such devices with high gain for a small size and low power optical pumping [1]. High doping concentrations are reachable in doped inorganic glasses or crystals, but the material should be processable as thin films in order to make planar optical waveguide amplifiers. Such device has been previously proposed [1], but sophisticated ion implantation techniques are needed. An alternative material would be organic erbium (III) doped materials [2]. They have numerous advantages, among them the possibility to tune the physico-chemical properties of the material by ligand formulation. Moreover, the use of absorbing ligand can allow the organic part of the complex to act as a photosensitizer, resulting in a several orders of magnitude increase in the absorption cross section of the material and a broader absorption band [3]. Such a light harvesting organic antenna could allow the use of low cost light emitting diode as the power source in a planar amplifier.

Most of organic materials behave as efficient NIR luminescence quenchers due to the coupling between the electronic transition of the erbium ion and the vibrationnal overtones of O-H, N-H and C-H bonds in the organic matrix or the ligand itself [2]. Such coupling is very efficient and luminescence lifetimes as short as a few microseconds are commonly observed. The consequences are a decrease of the emission quantum yield and the luminescence lifetime, which is up to several milliseconds in an inorganic medium. The substitution of hydrogen atoms by fluorine on the organic ligand is usually proposed and can be effective [4]. Moreover, fluorinated ligand confer to the complex volatility under reduced pressure allowing the complex to be sublimated.

We report here the use of a vacuum deposition technique to obtain a pure organolanthanide chelate thin film. This technique allows the film to be deposited on any substrate resulting in the highest erbium doping concentration reachable using this class of material. This study is aimed to the fabrication, the characterization of a vacuum deposited erbium doped organic thin film and the determination of the maximum concentration that can be reached in such a material. We use a previously reported structure based on the commercial erbium tris(hexafluoroacetylacetonate) complex with a tris(pentafluorophenyl)phosphine oxide coligand (TPPO) added to complete the coordination sphere of the complex [5] (Figure 1). Moreover the use of TPPO could help to limit the presence of water in the coordination sphere.

EXPERIMENTAL DETAILS

All reactions were carried out under inert nitrogen atmosphere unless otherwise specified. All the chemicals and others solvents from Aldrich were used as received unless otherwise specified. The erbium tris(hexafluoroacetylacetonate) from Strem Chemicals was used without further purification. ^{19}F and ^{31}P nuclear magnetic resonance (NMR) measurements were recorded on a Bruker AC-300 spectrometer. FTIR spectra were recorded with a Thermo-Nicolet Nexus spectrometer. The UV-visible spectra were recorded with a Varian Cary 5000 spectrometer. The melting point was measured with a DSC-7 apparatus from Perkin-Elmer at a heating rate of $20°C.min^{-1}$. For time resolved luminescence, the excitation source is a μJ to mJ output ~7 ns temporal FWHM optical parametric oscillator (OPO) with a frequency tripled YAG as the pumping source. A high speed 300 μm diameter pin InGaAs photodiode, amplified with a high speed (BW from $\leq$ 250 Hz to 60 MHz) low noise (NEP = 2.5 pW/Hz$^{1/2}$) amplifier, is used for detection of the luminescence. Thin films were deposited using a Edwards Auto 306 vacuum deposition chamber. Spectroscopic ellipsometry measurements were carried out using a Jobin-Yvon UVISEL apparatus. X-ray experiments are performed on a Bruker-AXS D8 Discover X-ray diffraction system provided with a seven-axis $(x-y-z-\chi-\varphi-(\xi-\zeta))$ precision sample stage and a vertical theta-theta goniometer, with a minimum step size of 0.0001degree. Cu (Kα) radiation at 1.54184 angströms is used in a Bragg-Brentano geometry. A graphite back curved monochromator is added to prevent the scintillation detector from sample secondary x-ray fluorescence and Compton radiation.

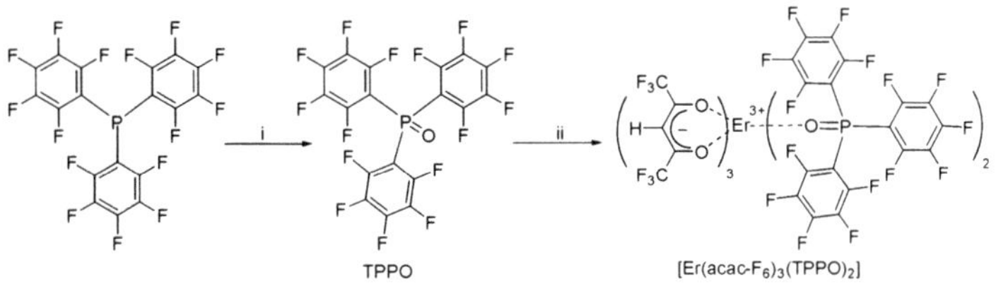

Figure 1. Synthesis of [Er(acac-F$_6$)$_3$(TPPO)$_2$] (i. H$_2$O$_2$ 33wt%, n-hexane/H$_2$O, 95%. ii. [Er(acac-F$_6$)$_3$, CH$_2$Cl$_2$, 91%)

Synthesis of tris(pentafluorophenyl)phosphine oxide TPPO [6]

To 40 mL (0,388 mol, 49 eq) of a 33 wt% hydrogen peroxide dissolved in 80 mL of deionized water is added drop wise at room temperature 4.2 g (7.9 mmol, 1 eq) of tris(pentafluorophenyl)phosphine in 240 mL of *n*-hexane solution. The solution is maintained 4 hours at 60-70°C under vigourous stirring. The solution is filtered and concentrated under vacuum. The residue is dissolved in dichloromethane and dried overnight on sodium sulphate. After filtration the solvent is evaporated. The crude is recrystallized from *n*-hexane to afford 4.11 g (7.5 mmol, 95%) of tris(pentafluorophenyl)phosphine oxide TPPO as white needles.
^{19}F NMR (300 MHz, CDCl$_3$, ppm, ref. C$_6$F$_6$) : δ -158.5 (2F, m), -142.6 (1F, m), -132.5 (2F, m). ^{31}P NMR (300 MHz, CDCl$_3$, ppm, ref. H$_3$PO$_4$ 85%) : δ -8.2 ppm; IR(cm^{-1}): 1648 (m), 1521 (m), 1489 (s), 1389 (m), 1301 (m), 1232 (m), 1098 (s), 985 (s), 768 (w), 731 (w); m.p. : 166°C.

Synthesis of [Er(acac-F$_6$)$_3$(TPPO)$_2$] complex [6]

To a solution of 1.96 g (2,48 mmol, 1 eq) of erbium tris(hexafluoroacetylacetonate) dissolved in 150 mL of dichloromethane is added 2.87 g (5,24 mmol, 2.11 eq) of tris(pentafluorophenyl) phosphine oxide. The solution is heated to reflux during 1 hour. The solution is cooled at room temperature and filtered. The filtrate is evaporated affording 4.39 g of crude product as a slightly pink powder. The crude is purified by vacuum sublimation at 180°C under 10^{-3} torr pressure.

RESULTS

The Figure 2 shows the absorption spectrum of a 20 g.L^{-1} (~10^{-2} mol.L^{-1}) solution of the [Er(acac-F$_6$)$_3$(TPPO)$_2$] complex in trichlorotrifluoroethane (TCTFE). The spectrum shows the absorption bands typical of the erbium ion. The study of the ^{4}I$_{13/2}$ absorption band around 1500 nm using different concentration TCTFE solutions allows the determination of the molar absorption coefficient of the complex and the calculation of the absorption cross section spectrum. A maximum molar extinction coefficient ε = 1.4 L.mol^{-1}.cm^{-1} is found at 1504 nm, corresponding to an absorption cross section of 5.2.10^{-21} cm². This value is coherent with the ones for other organic systems reported in the literature [7]. As a comparison it can be stressed that the organic moiety of the complex shows a molar absorption coefficient above 30 000 L.mol^{-1}.cm^{-1} in the UV part of the spectrum (300 nm), thereby confirming the interest in the use of ligand to metal energy transfer for an efficient pumping of the erbium ion.

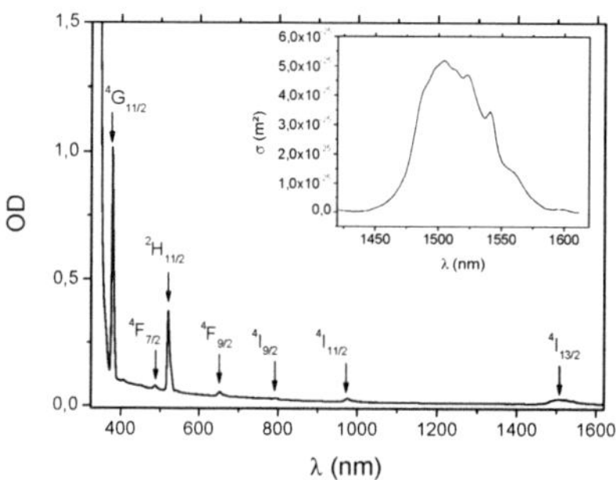

Figure 2. UV-vis-NIR absorption spectrum of a 20 g.L^{-1} (~10^{-2} mol.L^{-1}) solution of the [Er(acac-F$_6$)$_3$(TPPO)$_2$] complex in trichlorotrifluoroethane and the calculated absorption cross section spectrum of the $^4I_{13/2}$ transition of the ion around 1.5 µm.

The [Er(acac-F$_6$)$_3$(TPPO)$_2$] complex is placed in a Ta boat and heated in a 5.10^{-7} torr pressure chamber. The deposition is made at an average deposition rate of 4 nm.s^{-1} on silicon and fused silica substrates. The Figure 3 shows the luminescence decay of a solid sample (powder) and of the vacuum deposited thin film on silica of the [Er(acac-F$_6$)$_3$(TPPO)$_2$] complex after excitation at 520 nm. The erbium ion is then directly excited using the $^2H_{11/2}$ level. Both curves are fitted with a single exponential decay giving a lifetime of about 3.2 and 3.1 µs for the powder and thin film sample respectively in good accordance with published values [5].

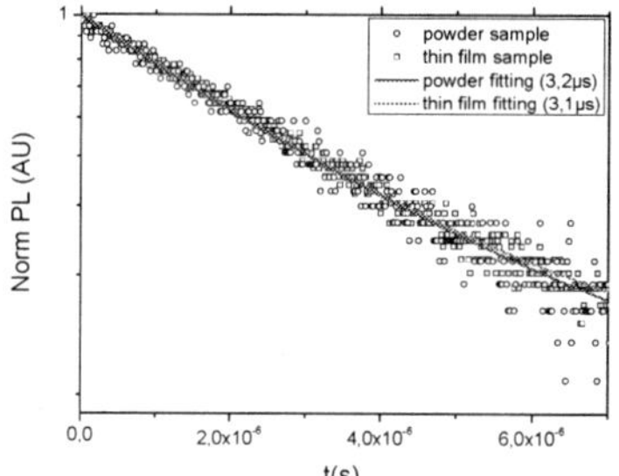

Figure 3. NIR (λ >1400 nm) luminescence decay of the [Er(acac-F$_6$)$_3$(TPPO)$_2$] complex in the form of powder and evaporated thin film on silica after short excitation at 520 nm.

The thin film on silica is observed by scanning electron microscopy (Figure 4a). The film is clearly visible and its thickness can be estimated to be around 725 nm. Moreover, a non quantitative EDS characterization confirms the presence of erbium and all the elements expected from the [Er(acac-F$_6$)$_3$(TPPO)$_2$] complex. Using ATR-FTIR (Figure 4b), the infrared absorption spectra of the complex powder and the thin film on silica can be compared. The thin film sample shows exactly the same spectrum which confirms that the complex has been sublimated under vacuum, has been deposited on the substrate and has not been deteriorated during the deposition process.

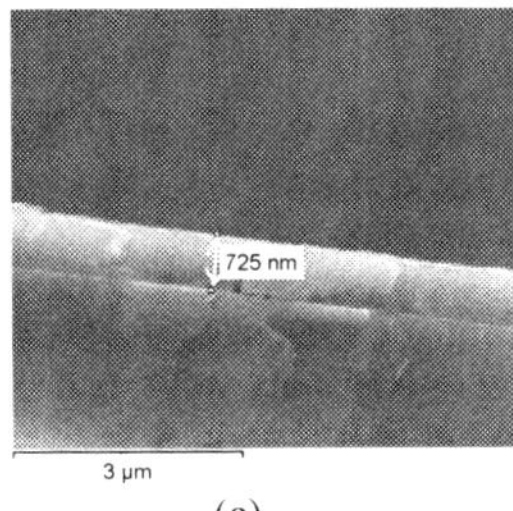

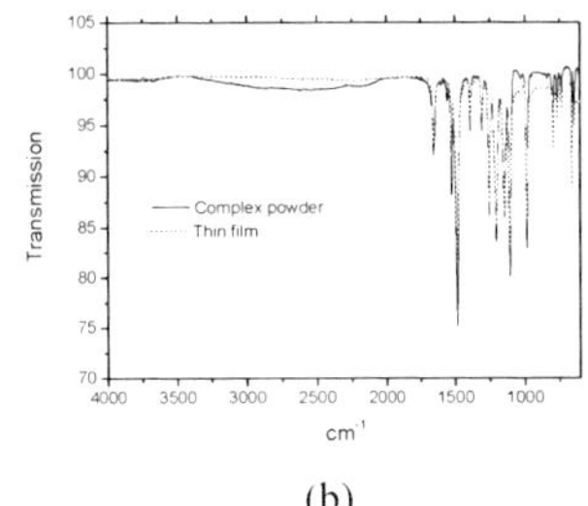

(a)   (b)

Figure 4. (a) SEM image of a thin film on silicon cleaved. (b) ATR-FTIR spectra of [Er(acac-F$_6$)$_3$(TPPO)$_2$] complex powder and thin film on fused silica substrate.

The thin film on silicon is studied using a spectroscopic ellipsometer (300-1700 nm). The ellipsometric spectra are fitted using a single oscillator model [8]. Due to the difference in absorption cross section between the organic moiety and the erbium ion, only the UV absorption band can be seen in the experimental data. The fitting of the experimental curves allows obtaining a film thickness of 778 nm. The refractive index at 1.5 µm is found to be ~1.46.

The Figure 5 shows a typical spectrum of a thin film sample on silicon. The spectrum shows only Bragg peaks on planes of polycrystalline phases, parallel to the surface. No amorphous phase is detected. The application of the Scherrer formula to the full width at half maximum (FWHM) of the deconvoluted peaks, using pseudo-Voigt functions, of figure 5 leads to a grain size determination ranging from 50 to 250 nm. The existence of a lot of peaks at low Bragg angles indicates that the crystallization occurs in a low symmetrical system. X-ray peak positions are closely related to data from literature [5] and structure refinement are in progress to extract the crystal system and lattice parameters.

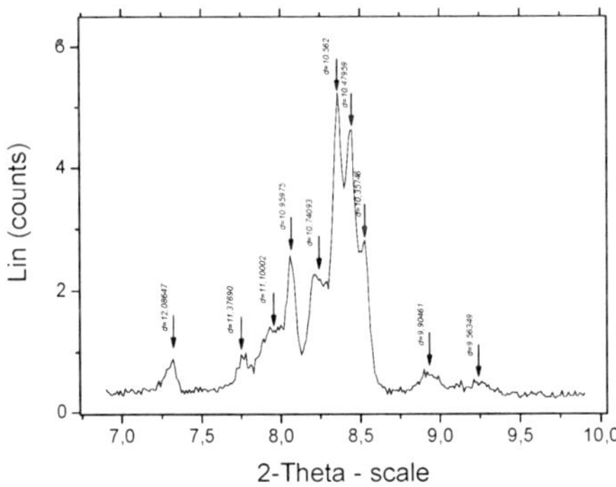

Figure 5. XRD spectrum of thin film of [Er(acac-F$_6$)$_3$(TPPO)$_2$] on silicon.

51

DISCUSSION

UV-Vis-NIR absorption spectrum measurements performed on the thin film with fused silica as substrate results in the absorptivity value around 1.5 μm. A $1.3.10^{-4}$ optical density is found and can be attributed to the erbium ion absorption. The emission lifetime of the erbium in the thin film can be determined the same way that in the powder sample. A 3.2 μs value is found for the luminescence lifetime of the thin film, in accordance with the value obtained for the complex powder. The luminescence lifetime of the thin film is similar to the published luminescence lifetime of a 10^{-5} M solution of the [Er(acac-F$_6$)$_3$(TPPO)$_2$] complex in anhydrous CCl$_4$ (τ = 2.4 μs, [5]). In the solid state, the concentration quenching seems to be negligible compared to the NIR luminescence quenching due to the coupling between the electronic transition of the erbium ion and the vibrationnal overtones of residual C-H bond of the hexafluoroacetylacetonate ligand.

Using the density of 2.027 g.cm^{-3} previously reported for a single crystalline sample of the same complex [5] and the theoretical weight percent of erbium ion in the complex (8.87%), a theoretical erbium ion concentration is calculated to be $6.5.10^{20}$ atoms.cm^{-3}.

Using the same density value, the absorption cross section determined from the solution of the complex, the thickness determined by spectroscopic ellipsometry and the optical density of the thin film at 1500 nm, an experimental concentration has been calculated to be $\sim 7.10^{20}$ atoms.cm^{-3}, in very good accordance with the theoretical value. Such an erbium concentration is interesting since it is a rather high concentration for luminescent erbium doped materials. Indeed typical organic materials are mostly constituted with a few percents of erbium chelate acting as dopant in a polymer film. Here the organic complex is present as crystallites with minimal organic content, resulting in the highest doping concentration for organic complex based materials. Such high doping concentration is interesting for compact planar waveguide amplifiers, in which high gains are needed.

CONCLUSION

We have reported the preparation and the characterization of organic thin films obtained with organic [Er(acac-F$_6$)$_3$(TPPO)$_2$] erbium complexes deposited by a vacuum sublimation technique on fused silica and silicon substrates. The deposited materials show high erbium concentration that could provide high gains in optical devices made from this material. The use of the vacuum sublimation technique is compatible with classical microelectronic processes allowing the fabrication of integrated planar waveguide amplifiers. The short excited state lifetime ($\tau \sim 3$ μs) is mainly due to the presence of a residual C-H bond in the coordination sphere of the Er ion and is probably too short for optical amplification to occur at reasonable pump powers. Efforts are now concentrated on the synthesis of perfluorinated erbium complexes to obtain thin films combining high erbium concentration and longer excited state lifetimes, keeping in mind that the excited state lifetimes of such perfluorinated erbium complexes thin films would be probably limited by the Er-Er quenching.

ACKNOWLEDGMENTS

The authors gratefully acknowledge the French National Research Agency (ANR) for the financial support through the VERSO program METAPHOTONIQUE.

REFERENCES

1. P.G. Kik and A. Polman, *MRS Bulletin* **23**, 48, (**1998**)
2. K. Kuriki, Y. Koike and Y. Okamoto, *Chem. Rev.* **102**, 2347 (**2002**)
3. L. H. Slooff, A. Polman, M. P. Oude Wolbers, F. C. J. M. van Veggel, D. N. Reinhoudt and J. W. Hofstraat, *J. Appl. Phys.* **83**, 497 (**1998**).
4. G. Mancino, A. J. Ferguson, A. Beeby, N. J. Long, T. S. Jones, *J. Am. Chem. Soc.*, **127**, 524 (**2005**)
5. A. Monguzzi, R. Tubino, F. Meinardi, A. Orbelli Birolli, M. Pizzoti, F. Demartin, F. Quochi, F. Cordella and M.A. Loi, *Chem. Mater.* **21**, 128 (**2009**)
6. L. Song, J. Hu, J. Wang, X. Liu and Z. Zhen, *Photochem. Photobiol. Sci.* **7**, 689 (**2008**)
7. A. Q. Le Quang, V. G. Truong, A.-M. Jurdyc, B. Jacquier, J. Zyss and I. Ledoux, *J. App. Phys* **101**, 023110 (**2007**)
8. G. E. Jellison, Jr. and F. A. Modine, *Appl. Phys. Lett.* **69**, 371 (**1996**)

Mater. Res. Soc. Symp. Proc. Vol. 1342 © 2011 Materials Research Society
DOI: 10.1557/opl.2011.863

Multicolor Luminescence from $Ca_3Y_2(SiO_4)_3$:Eu^{2+},Eu^{3+} Material

Anna Dobrowolska and Eugeniusz Zych

Faculty of Chemistry, University of Wroclaw

14 F. Joliot-Curie Street

50-383 Wroclaw, Poland

ABSTRACT

$Ca_3Y_2(SiO_4)_3$ powder activated with Eu was prepared with ceramic method. Its spectroscopic properties in VUV-UV-Vis region together with structure and morphology were analyzed. Luminescence measurements indicated that $Ca_3Y_2(SiO_4)_3$ prepared in a reducing atmosphere contained both Eu^{2+} and Eu^{3+} ions. The excitation could be tuned to generate a complex luminescence consisting of a broad band related to a blue-greenish $5d \rightarrow 4f$ emission of Eu^{2+} and $4f \rightarrow 4f$ red luminescence of Eu^{3+}. Their superposition covered the whole visible part of spectrum and appeared almost white among others upon excitation around 390 nm.

INTRODUCTION

Light emitting diodes (LEDs) are one of the most rapidly evolving branches of the solid state lighting technology in recent years. Special interest is devoted to white LEDs due to their high brightness and efficiency, long lifetime and low power consumption [1]. To make white LEDs, blue or UV emitting LEDs are combined with phosphor(s) able to convert the radiation of a semiconductor chip into longer-wavelengths visible light [2, 3, 4]. Blue and UV-LEDs require different phosphors. Especially for the latter a mixture of phosphors is usually needed to cover with their luminescence possibly entire range of the visible part of spectrum. Our goal is to design a single phosphor efficiently stimulated with a near UV-LED (340-400 nm) and able to produce white light with spectral characteristics similar to the natural sunlight. At first, to make such a material we explored $Ca_3Y_2(SiO_4)_3$ host doped with two different europium ions, Eu^{2+} and Eu^{3+}. We considered this host attractive as it offers metal sites of both +2 (Ca) and +3 (Y) charges. We anticipated that europium, under favorable conditions of the material fabrication, should tend to enter the host as to distinct luminescent centers: a broad band emitting Eu^{2+}, whose $5d \rightarrow 4f$ luminescence might well locate in blue-green part of spectrum, and Eu^{3+} which sends off red photons due to the 4f $\rightarrow$ 4f transitions. Till now this lattice was explored as a host for trivalent lanthanides only [5, 6, 7, 8]. We shall present preliminary results on luminescent spectroscopy of the $Ca_3Y_2(SiO_4)_3$:Eu^{2+},Eu^{3+} phosphor and will demonstrate that introducing only one element as a dopant (Eu) and applying precisely defined synthesis conditions may be enough to get efficient white light emitting phosphor.

EXPERIMENT

Eu-doped calcium yttrium silicate, $Ca_3Y_2(SiO_4)_3$, was synthesized with the classic ceramic method calcining a mixture of CaO (99.6-100.5 %), Y_2O_3 and (99.99 %), SiO_2 (AR) and Eu_2O_3 (99.99 %) powders. Eu was assumed to substitute Y^{3+} exclusively giving $Ca_3Y_{2(1-x)}Eu_{2x}(SiO_4)_3$, where x= 0.01. The substrates were thoroughly ground. After drying the mixture was heated twice in the reducing atmosphere of H_2 (25 %)-N_2 (75%) mixture at 1400 °C for 6 h with an intermediate grinding. Photoluminescence and excitation spectra were recorded with 0.2 nm resolution using an FSL 920 spectrometer from Edinburgh Instruments. The excitation spectra were recorded with the emission monochromator slits set to 2 nm. A 450 W Xe-lamp was used as an excitation source, and both types of spectra were corrected for the system characteristics. The luminescence and luminescence excitation spectra were also recorded with synchrotron radiation at the Superlumi station of DESY-Hasylab in Hamburg, Germany. These emission spectra were recorded with a CCD camera with the resolution of about 0.25 nm. The synchrotron excitation spectra were corrected for the incident light intensity using sodium salicylate as a standard and the emissions were corrected for the spectral characteristics of the detection system. TEM images were taken with FEI Tecnai G^2 20 X-TWIN transmission electron microscope. The X-ray diffraction (XRD) patterns were measured with a D8 Advance diffractometer (BRUKER) using Cu Kα1 radiation (λ=1.540596Å) in the range of 2θ=5-90 degrees with the $\Delta\theta$=0.016 step and the counting time of 0.5 s. FTIR spectrum was recorded with an IFS 66/s Bruker spectrometer in the range of 400-4000 cm^{-1}. KBr pallets were used.

DISCUSSION

Structure and Morphology

The measured X-ray diffraction pattern, presented in Fig. 1a, matches the one expected for orthorhombic system, space group Pnma according to PDF-2 database, file #01-087-0453 [9,

10]. Hence, the XRD confirmed crystallographic purity of the product. TEM images, presented in Fig. 1b,c, reveal that crystallites making up the powder have very irregular shapes, and their sizes vary between 20-500 nm, roughly. Most of crystallites are agglomerated into grains up to ~1 µm in diameter. FTIR spectrum of Eu-doped $Ca_3Y_2(SiO_4)_3$, not presented here, consists of components around 500 cm^{-1} attributed to Ca/Y-O bond stretching mode vibrations and more intense, around 940 cm^{-1} assigned to Si-O-Si bond bending and Si-O stretching vibrations [11]. Hence, the vibrations are not so much energetic to promote efficient nonradiative de-excitation. This situation is similar as in the YAG lattice known to serve as a host for efficient phosphors.

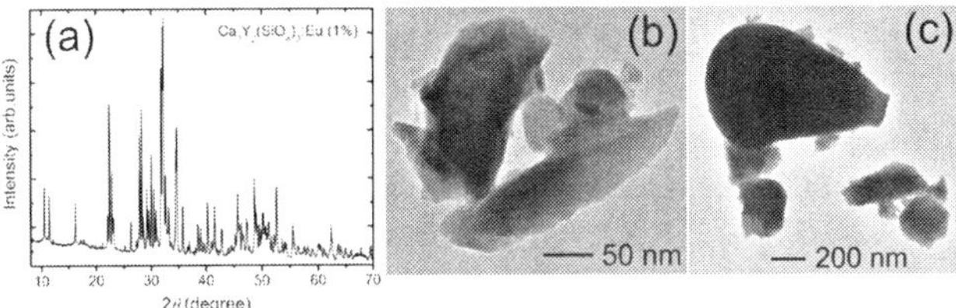

Figure 1. XRD pattern (a) and TEM images of $Ca_3Y_2(SiO_4)_3$:Eu powders (b-c).

Room temperature luminescence under UV excitation

Figs. 2a and 2b present luminescence excitation spectra of Eu^{3+} (614 nm) and Eu^{2+} (510 nm) emissions, respectively. The former spectrum is composed of broad component with maximum around 270 nm which should be assigned to charge transfer transition between O^{2-} and Eu^{3+} and line-type excitation features at longer wavelengths related to absorption within the $4f^7$ shell of Eu^{3+} [12, 13]. Their assignment is given in the Figure 2a [14]. The excitation spectrum of the broad emission peaking around 510 nm (Eu^{2+}) consists of two overlapping components with maxima about 280 nm and 350 nm. Both seem to appear due to allowed $4f \rightarrow 5d$ transition within Eu^{2+} ions [13]. It is interesting that around 392 nm there is a clear dip in the excitation spectrum of the 510 nm luminescence. Undeniably, it comes from the relatively strong competing absorption of Eu^{3+} at this wavelength (see Fig. 2a, the transition to $^5L_6$). Hence, the dip confirms that energy of the 392 nm radiation hitting the phosphor is being shared between both Eu^{2+} and Eu^{3+} and both these luminescent centers are being excited then.

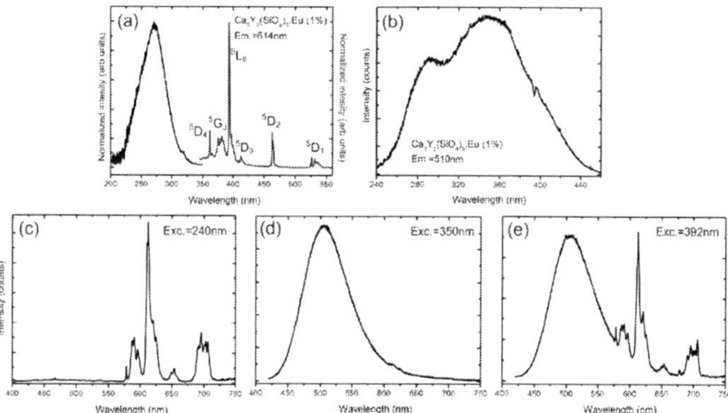

Figure 2. Luminescence excitation spectra of 614 nm (a) and 510 nm (b) emissions. Emission spectra of $Ca_3Y_2(SiO_4)_3$:Eu (1%) under different wavelengths of stimulating radiation: 240 nm (c), 350 nm (d) and 392 nm (e).

Both luminescence excitation spectra expose the range of useful excitation wavelengths to create luminescence in this phosphor. A high-energy stimulation around 240 nm leads to the red Eu^{3+} luminescence containing all characteristic features of intraconfigurational 4f → 4f transitions (Fig. 2c). Then, moving the excitation into longer wavelengths around 350 nm yields only a 5d → 4f broad band emission of Eu^{2+} (Fig. 2d). Nevertheless, setting the excitation into yet longer wavelengths around 392 nm (25510 cm^{-1}) produces luminescence spectrum containing both the Eu^{3+} and Eu^{2+} emissions superimposed (Fig. 2e). Consequently, the 392 nm excited luminescence appears white with characteristics not very much different than sunlight.

However, since the Eu^{3+} luminescence excitation spectrum around 390-400 nm consists of narrow lines it is rather sensitive to variations in stimulating wavelengths. Hence, the ratio of Eu^{2+} and Eu^{3+} emissions can also vary noticeably if the incoming radiation wavelength changes. This imposes rather strict requirements on the emission of LED coupled with the phosphor and supposed to stimulate it. This might be problematic especially in the case of high power LEDs, whose emission position may change to some extent with temperature. Fortunately, the repetitive synthesis and emission measurements proved that applying analogous fabrication conditions allows getting almost identical luminescence spectra. Hence, the ratio of Eu^{2+} and Eu^{3+} emitting ions could be effectively controlled through fabrication parameters. This is a good prognostic for further development of this phosphor.

<u>**Low temperature luminescence under VUV-UV synchrotron radiation excitation**</u>

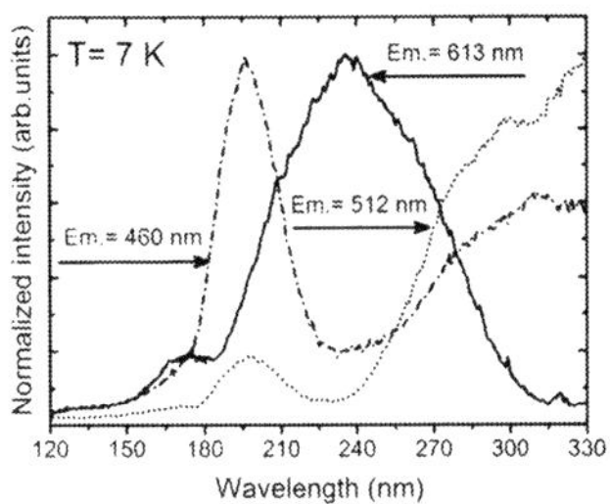

Figure 3. Excitation spectra of $Ca_3Y_2(SiO_4)_3$:Eu (1%) emission features positioned around 460 nm (mainly a defect), 512 nm (mainly Eu^{2+}) and 613 nm (Eu^{3+}) measured at 7 K.

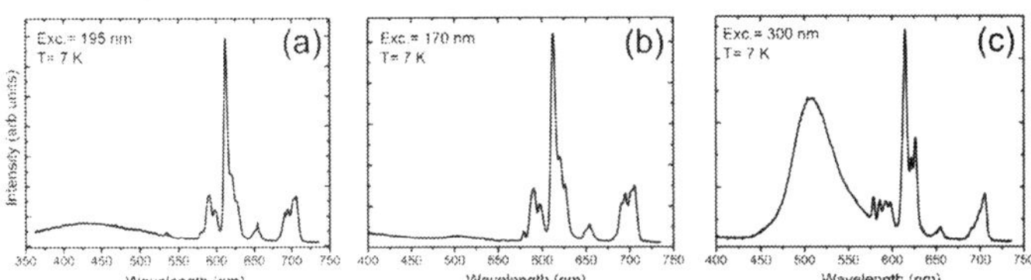

Figure 4. 7K luminescence spectra of $Ca_3Y_2(SiO_4)_3$:Eu (1%) powder recorded upon excitation at 195 nm (a), 170 nm (b), and 300 nm (c).

Luminescence excitation spectra of various luminescent features (460 nm, 613 nm and 512 nm) in $Ca_3Y_2(SiO_4)_3$:Eu (1%) recorded at the temperature of 7 K are presented in Figure 3. Excitation spectrum of Eu^{3+} luminescence located at 613 nm is composed of two bands peaking at 170 nm and 240 nm. The former could be interpreted as connected with the fundamental absorption of the host lattice and the latter as a CT band of Eu^{3+}. Excitation spectrum of Eu^{2+} luminescence positioned around 512 nm also contains two broad features, one around 195 nm and another one extending from 240 nm to beyond 330 nm, the limit of the available wavelengths at the Superlumi experimental station. The latter is assigned to $4f \rightarrow 5d$ transition of Eu^{2+} ions. A broad component appearing around 195 nm could be related to the defect absorption. This hypothesis is justified by 195 nm excited emission. Figure 4a shows that stimulation with 195 nm wavelengths, despite the red Eu^{3+} luminescence, produces a very broad component exceeding from 350 nm to 550 nm - quite characteristic feature of defect emission. The excitation spectrum of this luminescence is dominated by the band around 195 nm. Figure 4b presents emission spectrum upon excitation into the host fundamental absorption at 170 nm.

The spectrum contains only Eu^{3+} luminescence which indicates that the excited host transfers the acquired energy to Eu^{3+} ions, basically omitting the reduced Eu^{2+} dopant. At 7 K luminescence of divalent europium appears only upon a direct stimulation into $4f \rightarrow 5d$ band (Figure 4c).

CONCLUSIONS

$Ca_3Y_2(SiO_4)_3$:Eu was synthesized with ceramic method in the reducing atmosphere of N_2-H_2 mixture. UV-Vis luminescence spectroscopy indicated that europium exists in the host lattice at two valence states, as Eu^{3+} and Eu^{2+}. It was shown that depending on the stimulation radiation wavelength the ions can be excited either separately or simultaneously around 392 nm. In the latter case efficient white luminescence is produced. VUV-UV synchrotron measurements reveled that at the temperature of 7 K a defect also contributes to luminescence of $Ca_3Y_2(SiO_4)_3$:Eu. It was also shown that excited host is able to transfer its energy only to Eu^{3+}, while Eu^{2+} luminescence appears only after direct stimulation into $4f \rightarrow 5d$ absorption of this ion.

ACKNOWLEDGMENTS

The research was supported by EU Program of Innovative Economy POIG.01.01.02-02-006/09. Partial support through DESY HasyLab Grant #II-20090289 EC and Minister of Science and Higher Education Grant # N N209 044839 are also acknowledged.

REFERENCES

1. W. Ki, J. Li, *J. Am. Chem. Soc.* **130**, 8114-8115 (2008).
2. A. M. Srivastava, H. A. Comanzo, U. S. Pat. No. 7,015,510 B2 (2006).
3. E. Radkov, A. A. Setlur, US 2007/0276606 A1 (2007).
4. S. Naum, W. H. Lo, C. R. Tsai, US 7,816,663 B2, 2010.
5. K.V. Ivanovskikh' A. Meijerink, F. Piccinelli, A. Speghini, E.I. Zinin, C. Ronda, M. Bettinelli, *J. Lumin.* **130**, 893-901 (2010).
6. F. Piccinelli, A. Speghini, G. Mariotto, L. Bovo, M. Bettinelli, *J. Rare Earth.* **27**, 555-559 (2009).
7. V. R. Bandi, Yung-Tang Nien, In-Gann Chen, *J. Appl. Phys.* **108**, 023111 (2010).
8. V. R. Bandi, Yung-Tang Nien, Tzung-Heng Lu, In-Gann Chen, *J. Am. Ceram. Soc.* **92**, 2953–2956 (2009).
9. PDF-2 database , Release 2010.
10. H. Yamane, T. Nagasawa, M. Shimada and T. Endo, *Acta Cryst. C* **53**, 1367-1369 (1997).
11. Z. Andric, R. Krsmanovic, M. Marinovic-Cincovic, T. Dramicanin, B. Secerov and M.D. Dramicanin, *Acta Phys. Pol.A* **5**, 969-974 (2007).
12. G. Blasse,*J. Chem. Phys.* 48YR **45** (7), 2356–2360 (1966).
13. G. Blasse, B. C. Grabmaier in *Luminescent Materials*, vol. 10. Springer-Verlag, Berlin, 1994.
14. M. Karbowiak, E. Zych, J. Hölsä, *J. Phys- Condens Mat.*, 50/YR **15**, 2169–2181 (2003).

Mater. Res. Soc. Symp. Proc. Vol. 1342 © 2011 Materials Research Society
DOI: 10.1557/opl.2011.996

Efficient near-infrared luminescence and energy transfer in Nd-Bi codoped zeolites

Zhenhua Bai[1], Minoru Fujii[1], Yuki Mori[2], Yuji Miwa[1], Minoru Mizuhata[2], Hong-Tao Sun[3], and Shinji Hayashi[1]

[1] Department of Electrical and Electronic Engineering, Graduate School of Engineering, Kobe University, Rokkodai, Nada, Kobe 657-8501, Japan

[2] Department of Chemical Science and Engineering, Graduate School of Engineering, Kobe University, Rokkodai, Nada, Kobe 657-8501, Japan

[3] International Center for Young Scientists (ICYS), National Institute for Materials Science (NIMS), 1-2-1 Sengen, Tsukuba-city, Ibaraki 305-0047, Japan

ABSTRACT

We prepare Nd-Bi codoped zeolites by a method consisting of a simple ion-exchange process and subsequent high-temperature annealing. The emission covers the range of 970~1450 nm, corresponding to the electronic transitions of Nd^{3+} ions and Bi-related active centers (BiRAC), respectively. The introduction of Bi distinctly broadens the excitation band of Nd^{3+} in the visible region, and the lifetime of Nd^{3+} reaches as long as 354 µs. In the zeolite matrix, Bi ions exist as BiRAC and Bi oxide agglomerates. The former one act as a sensitizer of Nd^{3+} ions, and the latter one act as a blockage to avoid the quenching effect of coordinated water, which enable Nd^{3+} ions to show efficient near-infrared (NIR) emission even the zeolites contain large amount of coordinated water. The excellent optical and structural properties make these NIR emitting nanoparticles promising in application as laser materials and biological probes.

INTRODUCTION

In recent years, much attention has been paid to the research of rare-earth ions doped materials for the potential applications in laser, optical telecommunication, and many other applications [1-6]. Among them, Nd^{3+} is one of the most investigated near-infrared (NIR) emitting rare-earth ions for two reasons. On one hand, Nd doped materials are commonly used as active media for solid state laser applications [2,3]. On the other hand, for in vivo biological imaging, the optical absorption and light scattering of human tissue are minimal in the NIR region [4]. Rare-earth ions, which are luminescent in this region, such as Nd^{3+} at 1064 nm, are ideal candidates for in vivo biological probes [4,5]. However, excitation efficiency of Nd^{3+} is very low due to its forbidden intra-4f transition [3]. Moreover, the intrinsically sharp and discontinuous absorption bands of Nd^{3+} require lasers as excitation source to realize some functional applications. Thus, it is an interesting topic to realize sensitization of Nd^{3+} by using active centers with broad absorption bands. Recently, it was found that "bismuth-related active centres (BiRAC)" can be exploited as sensitizers of rare earth ions such as Er^{3+} and Yb^{3+} [6,7]. Therefore, we also expect that there may be effective sensitization of Nd^{3+} ions by an energy transfer process from BiRAC.

Zeolites are microporous crystalline aluminosilicates with nanosized pores. Their framework is composed of SiO_4 and AlO_4 tetrahedra units by sharing oxygen between every two consecutive units, and cations located inside channels or cavities to balance negative charges in the framework. Recently, zeolites are of great interest in the application as host materials of optically active guests, because the well-defined and well-organized cavities provided by zeolites host serve as an ideal environment to organize the optically active guests well dispersed in their framework [8-13]. However, it is difficult to realize highly efficient NIR emission in zeolites, because of the existence of coordinated water in cages, causing fast relaxation of its excitation energy through nonradiative vibrational deactivation [8,12]. Thus, the effective method to increase NIR emission efficiency in zeolites is to separate active ions from coordinated water.

In this work, we realize strong and long-lived Nd^{3+} emission in zeolites by means of codoping bismuth. The emission covers the range of 970~1450 nm, corresponding to the electronic transitions of Nd^{3+} ions and BiRAC, respectively. The NIR emission of Nd^{3+} ions is significantly enhanced by the introduction of bismuth in codoped samples, and the lifetime reaches as long as 354 μs. Steady state and time-resolved photoluminescence (PL), and PL excitation (PLE) measurements demonstrate the energy transfer from BiRAC to Nd^{3+}.

EXPERIMENTAL

The NH_4 form of faujasite- (FAU) type zeolite was purchased from Tosoh Co. Japan (Zeolite Y, SiO_2/Al_2O_3=7, grain size 700~1000 nm). Zeolites were stirred in x mM (x=0.5, 3, 6) aqueous solution of Nd^{3+} prepared from $Nd(NO_3)_3 \cdot 6H_2O$ at 80 °C for 96 h to exchange NH_4 ions with Nd^{3+} ions. The products were removed by centrifugation, then washed with deionized water, and dried in air at 120 °C. The Nd embedded zeolites were further stirred in a 60 mM aqueous solution of Bi^{3+} prepared from $Bi(NO_3)_3 \cdot 5H_2O$ at 80 °C for 120 h to dope Bi. The products were removed by centrifugation, then washed with deionized water, and dried in air at 120 °C. The Nd-Bi codoped zeolites were calcined at 920 °C for 1 h in N_2 atmospheric condition. Hereafter, the codoped samples were denoted as Z-xNd-yBi, where x and y are neodymium and bismuth concentrations (mM) in the experimental solution. For comparison, we also prepared Z-3Nd and Z-60Bi singly doped zeolites.

The structural properties of prepared products were characterized by an X-ray diffractometer (Rigaku-TTR/S2, λ=1.54056 Å) and a field emission scanning electron microscopy (FE-SEM). Neodymium and bismuth concentrations were determined by energy-dispersive X-ray spectroscopy (EDS). The atomic concentrations of Nd and Bi in the final products were shown in Table I . PL and PLE spectra were measured by using an Ar^+ laser and an optical parametric oscillator (OPO) pumped by the third harmonic of a Nd:YAG laser. The signal was analyzed by a single grating monochromator and detected by a liquid-nitrogen-cooled InGaAs detector. Time-resolved luminescence measurements were performed by detecting the modulated luminescence signal with a photomultiplier tube, and then analyzing the signal with a photon-counting multichannel scaler. The excitation source for the lifetime measurements was 530 nm light from the OPO (pulse width 5 ns, repetition frequency 20Hz). All the measurements were carried out at room temperature.

Table I . The atomic concentrations of Nd and Bi in the final products

Sample	Nd (at. %)	Bi (at. %)
Z-3Nd	0.21	-
Z-0.5Nd-60Bi	0.04	1.28
Z-3Nd-60Bi	0.17	1.37
Z-6Nd-60Bi	0.28	1.34
Z-60Bi	-	1.30

RESULTS AND DISCUSSION

Figure 1 shows the XRD patterns of Z-3Nd, Z-60Bi, and Z-3Nd-60Bi samples. The data of the diffraction peaks agree well with the standard values for the Zeolite Y (JCPDS No. 45-0112), indicating that the obtained samples are single phase and the codoped neodymium and bismuth do not cause any significant change. The FE-SEM images of corresponding samples are shown in figure 2 (a)-(c). The morphology and monodispersity of these samples remain almost unchanged.

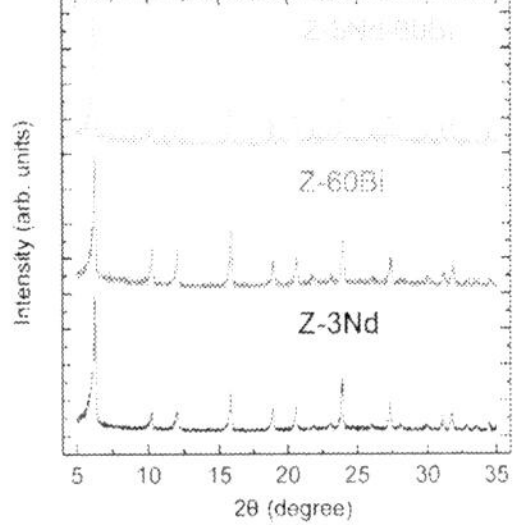

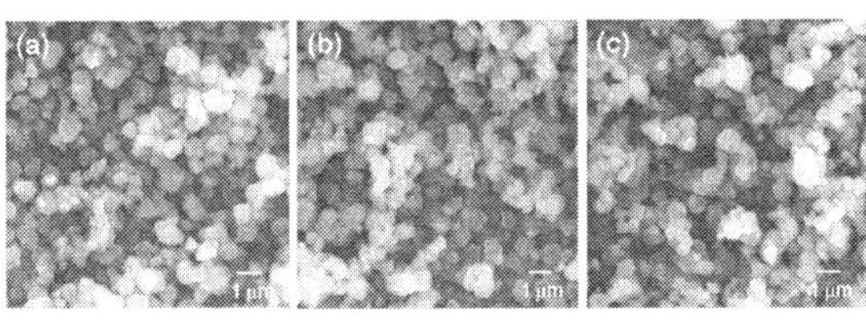

Figure 1. X-ray diffraction patterns of Z-3Nd, Z-60Bi and Z-3Nd-60Bi samples.

Figure 2. FE-SEM images of (a) Z-3Nd, (b) Z-60Bi and (c) Z-3Nd-60Bi samples.

Figure 3 shows PL spectra of Nd, Bi singly and codoped samples excited at 530 nm. In Bi singly doped sample, a broad NIR emission centered at 1145 nm is observed, which could be ascribed to BiRAC. The broadband emission covers the range of 970~1450 nm, and the full-width at half maximum is about 162 nm. For Nd singly doped zeolites, weak PL peaks can be seen at 1064 nm and 1332 nm due to the intra-4f shell electronic transitions ($^4F_{3/2} \rightarrow {}^4I_{11/2} + {}^4I_{13/2}$) of Nd^{3+} when it is excited at 530 nm. This wavelength can directly excite Nd^{3+} from the ground state to the upper excited state. In the codoped samples, strong emissions from Nd^{3+} and BiRAC are observed simultaneously. It can be seen that the Nd^{3+} emission in the codoped samples are much stronger than that of the singly doped sample, more than two orders of magnitude. It is interesting to note that, in the series of codoped samples, with the increase of Nd concentration, the emission intensity of BiRAC monotonically decreases compared with that in Bi singly doped sample, while the Nd^{3+} emission intensity increases. This indicates that the energy transfer from BiRAC to Nd^{3+} results in the gradual decrease of emission from BiRAC.

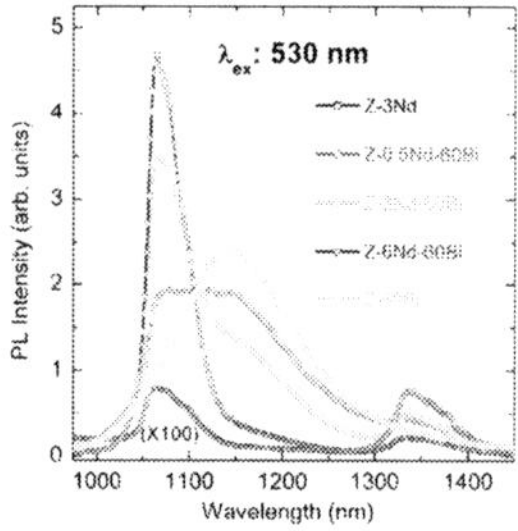

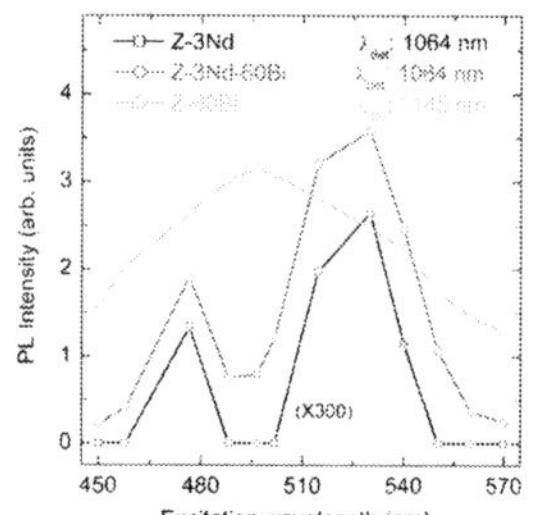

Figure 3. PL spectra of Nd, Bi singly and codoped zeolites excited at 530 nm.

Figure 4. PLE spectra of Z-3Nd and Z-3Nd-60Bi samples detected at 1064 nm, and Z-60Bi sample detected at 1145 nm.

In order to further investigate the energy transfer between BiRAC and Nd^{3+}, we studied excitation wavelength dependence of the 1064 nm peak. The measurements are performed under the same testing condition. The PLE spectra of Z-3Nd and Z-3Nd-60Bi samples detected at 1064 nm, and Z-60Bi sample detected at 1145 nm are shown in figure 4. The PLE spectrum of Nd singly doped sample displays two discrete excitation bands at 476 and 530 nm , corresponding to the $^{4}I_{9/2}$ to $^{2}K_{15/2}+^{2}D_{3/2}+^{2}G_{9/2}+^{4}G_{11/2}$ and $^{2}K_{13/2}+^{4}G_{7/2}+^{4}G_{9/2}$ transitions, respectively [14]. It is noteworthy that no Nd^{3+} emission can be detected under other excitation wavelengths. In contrast, the codoped sample shows a continuous excitation spectrum, which means that Nd^{3+} ions in the codoped sample are not only excited by direct absorption, but also excited by an indirect process. The indirect excitation of Nd^{3+} ions in the codoped sample indicates that an energy transfer process occurs in zeolites. On the other hand, the Bi singly doped sample exhibits a broad excitation band centred at ~500 nm (in figure 4), which coincides with those reported for Bi-doped glasses [15]. Therefore, BiRAC are most likely to act as a sensitizer of Nd^{3+}.

Figure 5(a) shows PL decay curves detected at 1064 nm. In the Nd singly doped sample, the lifetime is much shorter than the transition intrinsic lifetime of Nd^{3+} (less than 1 µs). This is due to nonradiative relaxation of Nd^{3+} by strong coupling of the excitation to high-frequency vibrations of coordinated water. The decay curves of codoped samples show a fast drop at the beginning, followed by a single-exponential slow component. With increasing Nd concentration, the slow component becomes more dominant. It is noteworthy that the Nd^{3+} emission at 1064 nm overlaps with the broadband emission of BiRAC, such that there are two components with different lifetimes in all curves. The occurrence of the fast drop in the decay curves is the result of relaxation of BiRAC excited state. The slow component is due to Nd^{3+} and the longest lifetime reaches 354 µs. The long lifetime of Nd^{3+} in codoped samples suggests that Nd^{3+} is well separated from coordinated water in zeolites. As is known, FAU type zeolites consist of three kinds of pores, the α-cages, the sodalite cages, and the hexagonal prisms. Nd^{3+} and Bi ions can migrate into the pores of zeolites by the ion-exchange process, and after subsequent high temperature annealing, Bi ions are converted to BiRAC and Bi compounds agglomerates, due to the low melting point of bismuth compounds. The agglomerates act as a blockage to seal the pores of zeolites, which can avoid the admission of coordinated water into the pores of zeolites. Therefore, Nd^{3+} ions can be held in captivity totally in the pores, and then have chance to show strong and long-lived NIR emission. The shortening of Nd^{3+} lifetime in the highest Nd concentration sample is due to the energy migration between neighboring Nd^{3+} ions.

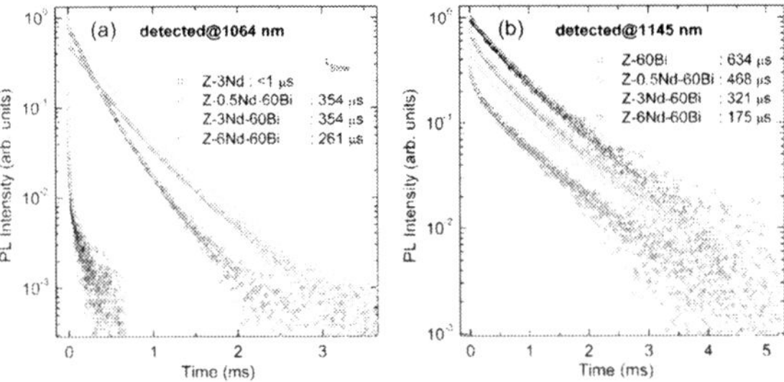

Figure 5. PL decay curves of all samples detected at (a) 1064 and (b) 1145 nm.

Figure 5(b) shows the decay curves detected at 1145 nm, i.e., emission from BiRAC. The decay curve of Bi singly doped sample can be well fitted by a single-exponential function, and the lifetime reaches as long as 634 µs. The curves of codoped samples can be fitted by double-exponential decay, and the mean lifetime (τ_m) was calculated by using equation 1.

$$\tau_m = \int_0^\infty [I(t)/I_0]d(t) \tag{1}$$

Where $I(t)$ is the luminescence intensity as a function of time t and I_0 is the maximum of $I(t)$ that occurs at the initial time t_0. The lifetimes of BiRAC decrease rapidly with the increase of Nd^{3+} concentrations, which is also a strong evidence of energy transfer from BiRAC to Nd^{3+}. Similar behavior is commonly observed for nonradiative energy transfer processes. The energy transfer efficiency η can be estimated by using equation 2.

$$\eta = 1 - \tau_{Nd-Bi}/\tau_{Bi} \tag{2}$$

Where τ_{Nd-Bi} and τ_{Bi} are the lifetimes of BiRAC in codoped and singly doped samples, respectively. The η is calculated to be 26 % (Z-0.5Nd-60Bi), 49 % (Z-3Nd -60Bi), and 72 % (Z-6Nd -60Bi).

For energy transfer process, it requires an overlap of the emission region of the sensitizer and the absorption region of the acceptor. The emission band of BiRAC covers the 970~1450 nm range, matching well with the excitation bands of Nd^{3+}, especially the $^4I_{11/2} \rightarrow ^4F_{3/2}$ (1064 nm) and $^4I_{13/2} \rightarrow ^4F_{3/2}$ (1332 nm) transitions. The results of measurements reveal that Nd^{3+} in the codoped system can be efficiently excited within the broad visible region, so optical pumping using flash lamp or white LED rather than laser becomes feasible using this broadband sensitization.

CONCLUSIONS

Nd-Bi codoped zeolites are prepared by an ion-exchange method and their optical properties are investigated. The emission covers the range of 970~1450 nm, corresponding to the electronic transitions of Nd^{3+} ions and BiRAC, respectively. The energy transfer between BiRAC and Nd^{3+} has been demonstrated from steady state and time-resolved emission spectra, and excitation spectra. The introduction of Bi distinctly broadens the excitation band of Nd^{3+} in the visible region. The energy transfer efficiency is also estimated. The excellent optical and structural properties make these NIR emitting nanoparticles promising in application as laser materials and biological probes.

ACKNOWLEDGEMENTS

This work is partly supported by a research grant from Nippon Sheet Glass Foundation for Materials Science and Engineering.

REFERENCES

1. M. Fujii, M. Yoshida, Y. Kanzawa, S. Hayashi, and K.Yamamoto, *Appl. Phys. Lett.* **71**, 1198 (1997).
2. K. Zou, H. Guo, M. Lu, W. Li, C. Hou, W. Wei, J. He, B. Peng, and X. Bin, *Opt. Express* **17**, 10001 (2009).
3. J. Meng, J. Li, Z. Shi, and K. Cheah, *Appl. Phys. Lett.* **93**, 221908 (2008).
4. M. D. Ward, *Coord. Chem. Rev.* **251**, 1663 (2007).
5. J. Zhang, P. D. Badger, S. J. Geib, and S. Petoud, *Angew. Chem. Int. Edn.* **44**, 2508 (2005).
6. Z. Bai, H. Sun, T. Hasegawa, M. Fujii, F. Shimaoka, Y. Miwa, M. Mizuhata, and S. Hayashi, *Opt. Lett.* **35**, 1926 (2010).
7. Z. Bai, H. Sun, M. Fujii, Y. Miwa, T. Hasegawa, M. Mizuhata, and S. Hayashi, *J. Phys. D: Appl. Phys.* **44**, 155101 (2011).
8. Y. Wada, T. Okubo, M. Ryo, T. Nakazawa, Y. Hasegawa, and S. Yanagida, *J. Am. Chem. Soc.* **122**, 8583 (2000).
9. J. Rocha, and L. D. Carlos, *Curr. Opin. Solid State Mater. Sci.* **7**, 199 (2003).
10. M. Lezhnina, F. Laeri, L. Benmouhadi, and U. Kynast, *Adv. Mater.* **18**, 280 (2006).
11. G. Calzaferri, and K. Lutkouskaya, *Photochem. Photobiol. Sci.* **7**, 879 (2008).
12. K. Binnemans, *Chem. Rev.* **109**, 4283 (2009).
13. H. Sun, A. Hosokawa, Y. Miwa, F. Shimaoka, M. Fujii, M. Mizuhata, S. Hayashi, and S. Deki, *Adv. Mater.* **21**, 3694 (2009).
14. Y. Chen, Y. Huang, M. Huang, R. Chen, and Z. Luo, *J. Am. Ceram. Soc.* **88**, 19 (2005).
15. Y. Fujimoto, and M. Nakatsuka, *Jpn. J. Appl. Phys.* **40**, 279 (2001).
16. H. Sun, Y. Sakka, Y. Miwa, N. Shirahata, M. Fujii, and H. Gao, *Appl. Phys. Lett.* **97**, 131908 (2010).

Mater. Res. Soc. Symp. Proc. Vol. 1342 © 2011 Materials Research Society
DOI: 10.1557/opl.2011.864

Red-emitting $Ca_{1-x}Sr_xS:Eu^{2+}$ Phosphors as Light Converters for Plant-growth Applications

Qi Xia[1,2], Miroslaw Batentschuk[1], Andres Osvet[1], Peter Richter[4], Donat P. Häder[4],
Jürgen Schneider[2], Lothar Wondraczek[3], Albrecht Winnacker[1] and Christoph J. Brabec[1]

[1]Chair WW6 Materials for Electronics and Energy Technology (i-MEET), University of
Erlangen-Nuremberg, Martensstr. 7, 91058 Erlangen, Germany
[2]Materials Research Center, University of Freiburg, 79104 Freiburg, Germany
[3]Institute of Glass and Ceramics (WW3), University of Erlangen-Nuremberg, Martensstr. 5,
91058 Erlangen, Germany
[4]Department Biology, Cell Biology Division, University of Erlangen-Nuremberg, Staudtstrasse
5, 91058 Erlangen, Germany

ABSTRACT

A series of $Ca_{1-x}Sr_xS:Eu^{2+}_{y\ mol\%}$ phosphors were synthesized with solid state reactions and
with various Ca/Sr ratio and Eu^{2+} doping concentrations. The influences of the lattice
composition and the Eu^{2+} doping level on photoluminescent properties were analyzed. With
doping concentrations between 0.1 to 3 mol%, concentration quenching takes place leading to
the decrease of luminance; the emission maxima are also red-shifted. Further, this work reports
enhanced photosynthetic activities of intact spinach leaves due to spectral modification of
simulated solar irradiation by one synthesized phosphor ($Ca_{0.4}Sr_{0.6}S:Eu_{1\ mol\%}$). The CO_2
assimilation rates of intact spinach leaves were monitored with an effective homemade
photosynthesis measurement system with controlled light conditions. The phosphor could
efficiently convert the photosynthetically less active green part of the solar spectrum into the red,
with a broad-band red emission centered at 650 nm and a halfband-width of 68 nm, giving an
excellent match with the absorption spectrum of spinach chloroplasts. By careful referencing the
photon flux, we found an enhanced photosynthetic activities by about 30 % due to the emission
of the phosphor.

INTRODUCTION

Solar irradiation serves as the ultimate energy source for photosynthetic activities of
higher plants. Green plants have selective utilization of light for their photosynthetic activities:
Chlorophyll *a* and *b* as the most important photosynthetic pigments show major absorption peaks
in the blue (about 430 nm and 450 nm, respectively) and red (640 nm and 660 nm, respectively)
spectral regions, thereby red photons are especially important for the photosynthesis [1]. Green
photons are less active for photosynthetic reactions and are mostly transmitted and reflected by
the leaves [1, 2]. An evident way to enhance the quantity of the red photons is to modify the
solar spectrum with a "green-to-red" converter. The emissive properties of $Ca_xSr_{1-x}S:Eu^{2+}$,
which was originally developed as spectral converter for LEDs [3], are matching the absorption
spectra of chlorophylls very well.

In our previous publication [3], the influence of host lattice compositions on
photoluminescent properties of the $Ca_xSr_{1-x}S:Eu^{2+}$ phosphors were investigated, revealing a
monotonic blue shift of emission maxima with higher Sr concentration, as well as a modest
thermal quenching upon increasing temperature up to 400 K. In the present work, we further
discuss the color tuning of red-emitting $Ca_{1-x}Sr_xS:Eu^{2+}$ as a sun light converter for plant
photosynthesis applications. An optimal Ca/Sr ratio is found providing a broad-band emission

centered at 650 nm which matches the absorption spectrum of chloroplasts of green plants. The doping content of Eu^{2+} also has an important influence on the quantum efficiency and emission maxima: concentrations between 0.3 mol% and 0.7 mol% provide highest luminance; further color tuning of the emission can be made by variations of the Eu content, however, a decrease of the luminance should be taken in account, especially when Eu concentration rises beyond 1 mol%. Finally, the results of enhanced photosynthetic activities of spinach are presented related to spectral modification of simulated solar light with one of the phosphors: $Ca_{0.4}Sr_{0.6}S:Eu_{1\ mol\%}$. The CO_2 assimilation rates of intact spinach leaves are monitored by a sensitive measurement system with controlled light conditions. Comparing to the reference experiments with MgO reflection foil, the converted light conditions lead to enhanced photosynthetic activities for about 30%. The improvement comes from the greatly enhanced photon numbers in the spectra range of $600 - 700$ nm due to the emission of $CaSrS:Eu^{2+}$ phosphor by green photon excitation.

EXPERIMENT

Preparation of photoluminescent materials

The series of $Ca_{1-x}Sr_xS:Eu^{2+}{}_y$ were prepared by sintering stoichiometric mixtures of the host materials (CaS/SrS), dopant source (Eu_2O_3) and flux NH_4Br at 1100°C for 2.5h in forming gas. The host lattice compositions were modified by increasing the SrS concentrations from 0 to 100% in the source powders. Dopant concentration of Eu^{2+} was varied from 0.1 to 3 mol% for all lattice compositions. To prepare the light conversion foil used later on in photosynthetic activity measurement system, the phosphor particles with an average size of 20 µm were obtained by post-treatments consisting of milling and sedimentation processes. They were then coated on a high reflective aluminum surface by the "Doctor Blade" method and were finally encapsulated in a transparent foil. With the same procedure the reflection foil was prepared with MgO (99.99 %, Alfa Aesar) particles.

Photosynthetic activity measurement system

Figure 1 shows the measurement system for monitoring the CO_2 assimilation of intact spinach leaves. The primary incident light was generated by four metal halide lamps (150 W) combined with green dichroic filters. Water shielding layer was placed above the cell to filter out the infrared spectral part from the lamps. The gas-tight glass cell (5 L) was totally blackened leaving a top light entrance. Intact spinach leaves were cut into six squares of 6 * 6 cm² each and fixed on a wet medium. The light spectra were measured by the spectrometer probe connected with a Spectrometer Carry 500, Varian. The CO_2 concentration, relative humidity and temperature inside were measured with gas sensors. For each measurement the starting CO_2 concentration was fixed as 443 ± 8 ppm by flushing compressed air through a water bottle, and CO_2 concentrations (in ppm) were logged every minute for 20 minutes; starting temperatures were about 22 °C and relative humidity was maintained above 90 %.

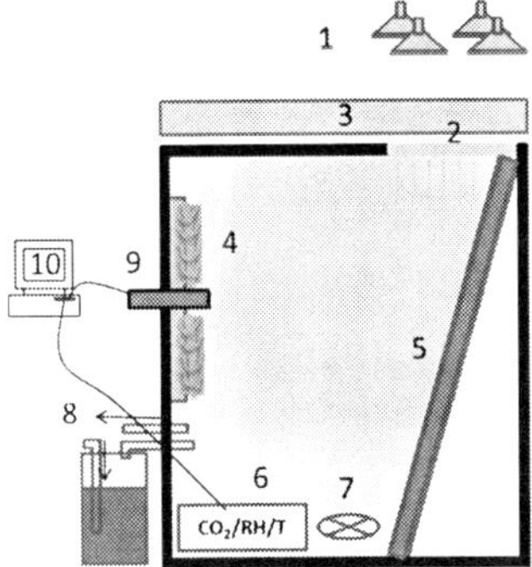

Figure 1. The photosynthetic activity measurement system consists of: 1 - metal halide lamps, 2 - dichroic filters, 3 - water shielding, 4 - intact leaves fixed on a sponge, 5 - conversion or reflection foil, 6 - CO_2/humidity/temperature sensors, 7 - ventilation fan, 8 – gas inlet/outlet for compressed air through a water bottle, 9 - spectrometer sensor probe, 10 – computer data logger.

DISCUSSION

Effect of Ca/Sr ratio on photoluminescent properties of the phosphors

As has been reported in our previous paper, the emission originating from the electric dipole $4f^65d \rightarrow 4f^7$ transition of Eu^{2+} shifts from 663 nm in CaS to 619 nm in SrS with fixed doping concentration of 1 mol% [3]. The observed shift is due to the fact that larger Sr-content results in a larger lattice constant and subsequently smaller crystal field splitting [4]. The proper Sr concentrations in the $Ca_{1-x}Sr_xS:Eu^{2+}_{1\,mol\%}$ phosphor could be therefore determined as x = 0.4 – 0.65 with the purpose to generate red light in the spectral range from 660 to 640 nm (s. (see Figure 2a) which is suitable for photosynthesis activities. Figure 2b compares the excitation spectra of $Ca_{1-x}Sr_xS:Eu^{2+}_{1\,mol\%}$ phosphors with various Sr concentrations.

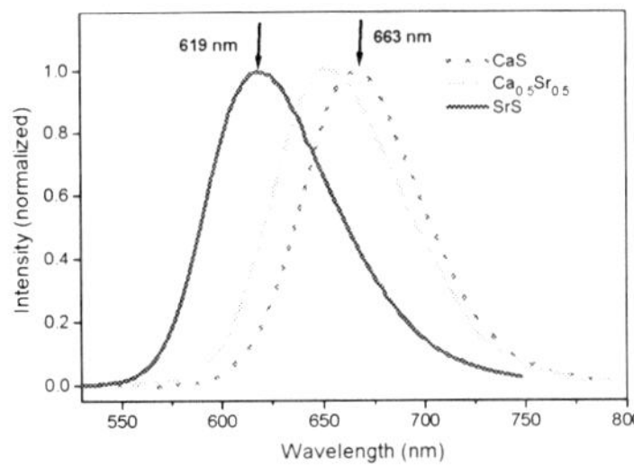
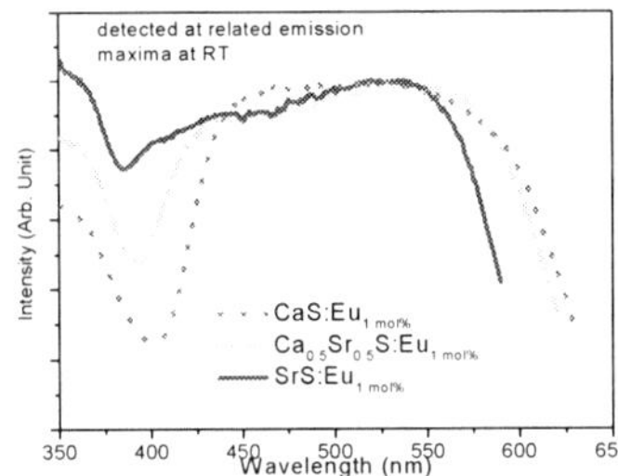

Figure 2. a - Normalized emission and b - excitation spectra of CaS, SrS and $Ca_{0.5}Sr_{0.5}S$ doped with 1 mol % Eu. The emission spectra were detected by excitation with a blue LED (460 nm); the excitation spectra were detected at related emission maxima.

Excitation spectra show broad bands in the visible region (440 - 560 nm) due to the characteristic $4f^7 \rightarrow 4f^6 5d$ transitions of Eu^{2+} in the cubic crystals [4]. It enables the phosphors to be efficiently excited by sun light rich in blue and green photons. Therefore the phosphors could be utilized as light converters for solar irradiation to supply tuned red irradiations for the photosynthetic activities of green plants.

Effect of Eu^{2+} concentration on photoluminescent properties of the phosphors

The influence of dopant concentration on PL properties was studied by increasing Eu^{2+} doping levels from 0.1 to 3 mol% in host lattices with identical composition - $Ca_{0.35}Sr_{0.65}S$. The emission spectra excited by a blue LED (460 nm) are presented in Figure 3a. The intensities and peak wavelengths are plotted against the dopant concentration in Figure 3b. The emission intensity increases with Eu concentration up to 0.3 mol% and decreases thereafter. When the concentration of dopant ions in the host lattice is sufficiently high, the excitation energy migrates due to the resonance between the activator ions, so that the energy eventually reaches the remote killers such as impurities, dislocations or the particle surface which act as quenching centers [4]. The existence of paired or even coagulated activator ions are more likely at higher Eu^{2+} doping levels, which may turn the activators into quenching centers. On the other hand, adequate doping level is necessary to achieve high absorption efficiency of the blue and green photons by the activator ions and hence high emission brightness. A balance needs to be reached for a high emission intensity with modest concentration quenching. Taking these considerations into account, the doping level of 0.3 - 0.7mol% is therefore considered as optimal.

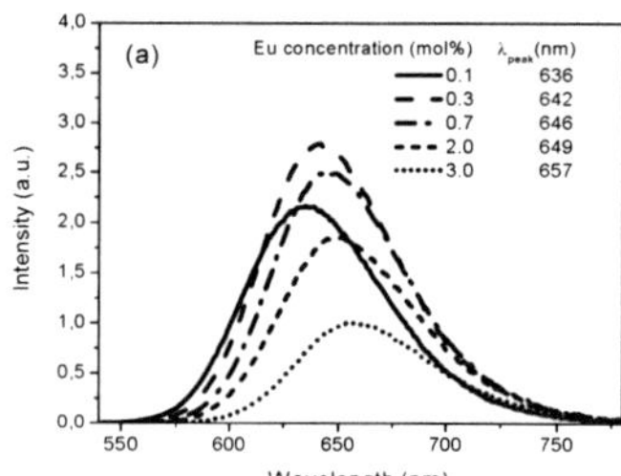
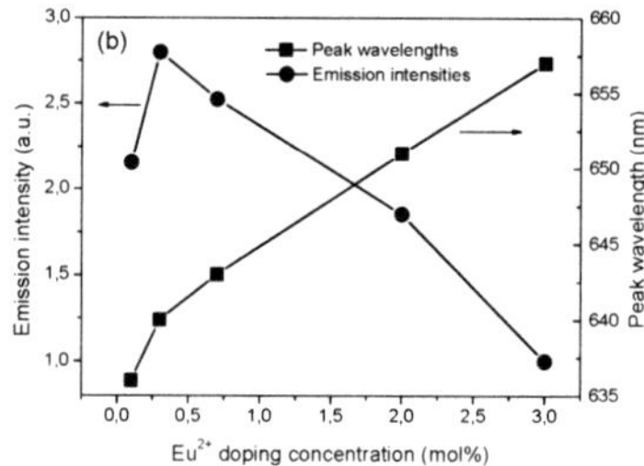

Figure 3. (a) Emission spectra of $Ca_{0.35}Sr_{0.65}S$ with various Eu^{2+} doping levels ranging from 0.1 to 3.0 mol% under the same excitation condition provided by a blue LED (460 nm); (b) emission intensities and peak wavelengths evaluated from Figure 3a at related dopant concentrations.

It is worth noting that a clear red shift upon increasing doping concentration is demonstrated in Figure 3b, similar to what was observed by increasing the Ca content in host lattices. We explain the shift with a stronger crystal field and a larger splitting of the 5d states of the Eu^{2+} ions surrounded by the relatively small Ca ions. The fraction of Ca-surrounded Eu^{2+} ions seems to rise with the Eu^{2+} concentration. Moreover, the shift can be explained partly by re-absorption of the blue wing emission band at higher concentrations.

<u>Spectral comparison of the modified/reference light conditions</u>

Figure 4 compares the spectra of light conditions when the reflection foil and the conversion foil are placed under the same primary incident light in the photosynthetic reaction cell. As shown by the dot line, the primary incident light is rich in green photons with wavelengths from 486 to 586 nm, and is reflected without spectral modification by the MgO surface. In comparison, when the conversion foil was applied, photon numbers in the red region are dramatically increased due to the emission of phosphor. The remaining green radiation results from the unabsorbed green photons as well as the reflection from the encapsulation foil. By quantitatively analyzing the spectra, green photons account for 95 % of the total photons in the unmodified conditions, while the photosynthetically more active blue and red photons are almost negligible. In the modified condition, the percentage of red photons is dramatically increased to more than 50 %, converted from the incoming green photons.

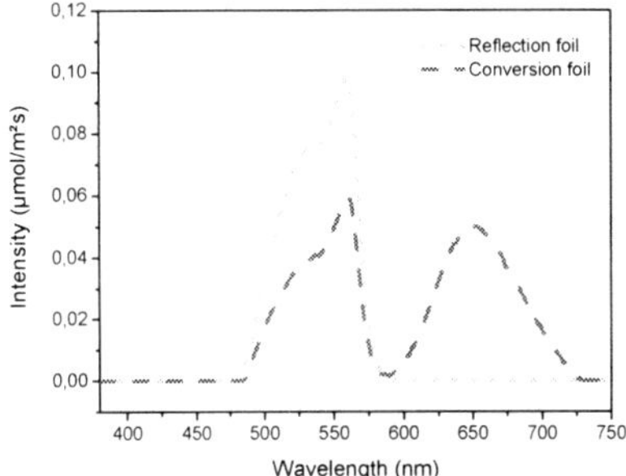

Figure 4. Spectra of light conditions provided by the reflective foil and the conversion foil under the same primary incident light measured in the photosynthetic reaction cell.

The CO_2 concentrations inside the biological cell were recorded with and without the spectral converter under identical primary incident light intensity of 1014.9 µmol/m² s (see Figure 5). In case of no spectral conversion, a highly reflective MgO mirror was used as reference. As it is known that photosynthesis activities are sensitive to temperature (T) and relative humidity (RH) [5,6], the environmental parameters were kept identical in the parallel experiments as shown by the inset: during each experiment procedure, the temperature increased slightly from a fixed starting temperature of about 22.1 °C to 23.8 °C; the RH remained relative constant above 90 %. From the CO_2 concentration curves, the decline rates were evaluated to be 11.0 and 13.9 ppm/min for the reference and modified conditions, respectively. The small magnitudes of error bars reveal the good reproducibility of the measurement system. Taking into account the following factors of the measurement system such as the total leaf surface area of 216 cm², and the cell volume of 5 L, one can calculate the specific CO_2 assimilation rates (µmol CO_2 molecules /m² s) for both conditions: 1.89 µmol/m² s for the reference, and 2.39 µmol/m² s for the conversion foil modified condition. Therefore the photosynthesis activity of spinach leaves in terms of CO_2 assimilation is increased by about 27 % as a result of light modification by the phosphor.

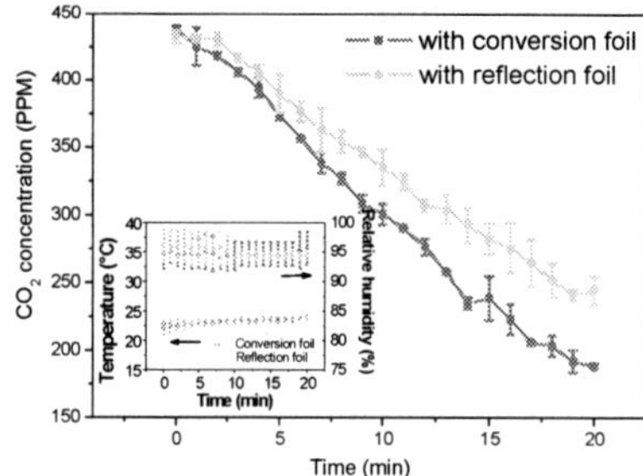

Figure 5. CO_2 concentration decline curves inside the reaction cell in the reference condition and the conversion foil modified condition under the same primary light intensity; inset: temperature and relative humidity during the measurements.

CONCLUSIONS

The influences of lattice composites and Eu^{2+} doping level on photoluminescent properties were analyzed. The influence of spectral modification of the light environment on photosynthetic activities of green plants were investigated as well. The CO_2 assimilation rates of intact spinach leaves were monitored using the developed photosynthesis measurement system with controlled light conditions. A conversion foil was prepared with imbedded $Ca_{0.4}Sr_{0.6}S$:Eu particles. The phosphor foil could efficiently convert photosynthetically less active green photons into active red photons, whose spectrum matches the absorption spectrum of spinach chloroplasts. Compared to the reference experiments with MgO reflection foils, the light conversion films did lead to significant enhanced photosynthetic activities with up to 30 %.

ACKNOWLEDGMENTS

The authors are grateful for the Bayerische Forschungsstiftung for kindly sponsoring the project (Project number: DOK-102-08). We also thank Dipl. Ing. S. Krolikowski, Dr. M. Peng, Dipl. Ing. A. Stiegelschmitt (WW3, FAU) for making characterization measurements and Dr. R. Auer, Dr. V. Kazuz, Dipl.-Phys. T. Swonke (ZAE Bayern) for providing the light facilities.

REFERENCES

1. L. Taiz, and E. Zeiger, *Plant Physiology*, 4th ed. (Sinauer Associates Inc., Sunderland, Massachusetts, 2006) p. 126-158.
2. D-P. Häder, *Photosynthese Taschenbuch*, (Thieme, Stuttgart, 1999).
3. Q. Xia, M. Batentschuk, A. Osvet, A. Winnacker, J. Schneider, Radiat. Meas. 45, 350 (2010).
4. S. Shionoya and W.M. Yen, *Phosphor Handbook*, (CRC Press Inc., 1998).
5. H.A. Mooney, C. Field, C.V. Yanes, C. Chu, Proc. Natl. Acad. Sci. USA. 80, 1295(1983).
6. G.E. Edwards and N.R. Baker, Photosynth. Res. 37, 89 (1993).

Mater. Res. Soc. Symp. Proc. Vol. 1342 © 2011 Materials Research Society
DOI: 10.1557/opl.2011.865

Photostimulable Fluorescent Nanoparticles for Biological Imaging

Andres Osvet[1], Moritz Milde[2], Sofia Dembski[2], Sabine Rupp[2], Carsten Gellermann[2], Miroslaw Batentschuk[1], Christoph J. Brabec[1] and Albrecht Winnacker[1]

[1]Chair Materials for Electronics and Energy Technology, University of Erlangen-Nuremberg, Martensstr. 7, 91058 Erlangen, Germany
[2]Fraunhofer Institute for Silicate Research, Neunerplatz 2, 97082 Wuerzburg, Germany

ABSTRACT

Spherical monodisperse core/shell-type nanoparticles, comprising an amorphous SiO_2 core coated with a luminescent phosphor layer were synthesized by the modified Pechini processes. The sol-gel method allows covering the 50 – 500 nm core particles with different inorganic phosphor layers of about 10 nm thickness, doped with rare-earth or transition metal ions which determine the luminescent properties. Particles comprising a Zn_2SiO_4 shell, doped with Mn^{2+} ions, are not only fluorescent under UV irradiation (260 nm), but store the activation energy by trapping electrons/holes at lattice defects. This energy is released as phosphorescence in the time scale of seconds and minutes, or as photostimulated luminescence under the excitation of red light (650 nm). Traps related to these processes are different, and their concentration is affected by the preparation conditions of the particles.

INTRODUCTION

Inorganic luminescent nanoparticles have many possible applications including phosphors for displays and conversion phosphors for white LEDs [1], transparent luminescent materials, or sensors [2]. Semiconductor quantum dots (QD) are considered as an alternative to the organic luminescent markers in biological and medical research and diagnostics [3], having broad absorption bands, less photobleaching and less reactions with nearby molecules. The problems related to QDs are the presence of toxic heavy metals, the luminescence intermittency, and the unwanted background luminescence created by the blue or UV excitation. Oxide- or fluoride-based nanoparticles may serve for marker applications as well; they have high chemical stability and brightness if doped with rare-earth (RE) or transition metal (TM) ions. Gated detection may be used to avoid the background emission, profiting from the long, micro- or millisecond decay time of the RE and TM ion emission. It is also possible to use infrared-excited upconversion luminescence like in $NaYF_4{:}Er^{3+},Yb^{3+}$ [4] with an additional advantage of the deep penetration of the infrared light and its smaller damaging potential to the cells. However, due to the low efficiency of the two-photon excitation, very high excitation intensity is needed.

In this work we focus on $Zn_2SiO_4{:}Mn^{2+}$ which belongs to phosphors in which the excitation energy may be stored by capturing the electrons/holes in traps. Afterglow over minutes, resulting from thermal releasing from the traps, opens further possibilities for marker applications as demonstrated in ref. [5]. Even more interesting is the possibility of triggering the luminescence of the previously charged particles at a desired moment, with only small thermal losses during the waiting time. The readout using red or infrared stimulating light may be complete, to maximize the signal, or partial, to enable "probing" of the samples repeatedly at different time intervals. In Mn-doped Zn_2SiO_4 the charging is done by irradiating the samples with UV light at 250 – 260 nm in the charge transfer band of the Mn^{2+} ions, or at shorter wavelengths over the bandgap of Zn_2SiO_4 [6]. The electrons are captured by different traps, contributing to phosphorescence and photostimulated luminescence (PSL). The bright green emission (λ_{max}= 525 nm) of Mn^{2+} is observed during the direct excitation and immediate relaxation of the ions, as an afterglow due to tunneling of the electrons back to the excited state of the Mn^{2+} ions, or during photostimulation by 650 nm light at a desired moment. The nature of the traps and their relation to the Mn^{2+} ions is not well understood, but the general idea is shown in figure 1, where the distance between the ground state of Mn^{2+} and conduction band is taken from ref. [7]. The aim of this work was to study the possibility of producing PSL in the corresponding nanoparticles. On one hand, the presence of the necessary lattice defects is

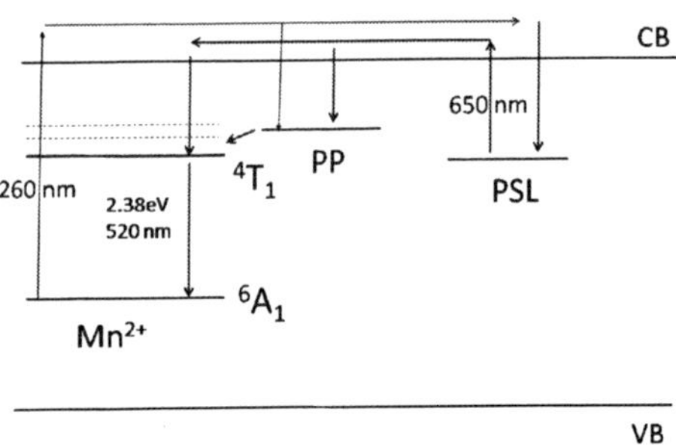

Figure 1. Tentative scheme of energy levels in Zn_2SiO_4:Mn^{2+}. PP and PSL designate the traps contributing to phosphorescence and photostimulated luminescence, respectively.

strongly dependent on the preparation method, temperature, and starting materials. On the other hand, the limited spatial separation between the trapped charges and the role of the surface may play an important role in nanoscale phosphors. Reports on phosphorescent $ZnMgSi_2O_6$:Mn^{2+}, Eu^{2+}, Dy^{3+} [5] and $SrAl_2O_4$:Eu,Dy [8] nanoparticles show the principal possibility at least in some systems. For the potential application it is important for the particles to be dispersable and of a definite size and morphology. For this reason core/shell SiO_2/Zn_2SiO_4 particles were prepared, where the size and morphology are controlled by the silica cores and the optical characteristics are determined by the phosphor coating.

EXPERIMENT

Silica cores of various size were prepared by the modified Stoeber method, the details of which have been published elsewhere [9,10]. Shortly, the TEOS was hydrolyzed in ethanol in the presence of ammonium hydroxide; the mixture was subsequently aged at RT. Before annealing the particles are covered with an organic framework containing the Zn and Mn ions. To this end isolated silica cores were added to the solution of the precursor salts of Zn^{2+} and Mn^{2+}. Citric acid was used to chelate the metal ions and PEG was the crosslinking agent. The resulting coated nanoparticles were collected by centrifugation, redispersed in water, freeze-dried and annealed at 800 to 1100 °C. Crystallinity of the powders was analyzed using a Philips PW1152.

The photoluminescence (PL) spectra were excited with a 75 Xe lamp through a 260 nm interference filter, and the spectra were recorded with a lock-in amplifier, R453 photomultiplier and HRS-2 monochromator. The PL decay was measured at the maximum of emission (525 nm) with a photomultiplier and a digital oscilloscope under the excitation of a pulsed Xe lamp through a 260 nm interference filter (4 µs pulses). The charging of the samples was done with the pulsed Xe lamp, at a pulse energy of 20 µJ/cm^2. Typical charging was done with 600 pulses. Phosphorescence decay was measured through a 520 nm interference filter with the photomultiplier by chopping the emission and using a lock-in amplifier. The PSL decay was recorded in the same way under the irradiation of the sample with a 650 nm LED, switched on after the phosphorescence had faded. The intensity of the stimulating light was 300 µW/cm^2. The phosphorescence spectrum was recorded with a monochromator and a CCD camera.

DISCUSSION

During the thermal treatment the organic framework is removed and a crystalline coating is formed: at temperatures below 900 °C mainly the formation of ZnO can be observed with the XRD analysis, at higher temperatures the α-willemite structure of Zn$_2$SiO$_4$ appears. The green photoluminescence, characteristic to the Mn^{2+} ions in this lattice appears simultaneously with the XRD lines. Annealing at 1100 °C produces a single phase of Zn$_2$SiO$_4$. The appearance of the pure phase of α-willemite, and the incorporation of the Mn^{2+} ions at regular tetrahedral lattice sites of Zn is confirmed by the photoluminescence measurements (fig. 2A). Under UV excitation

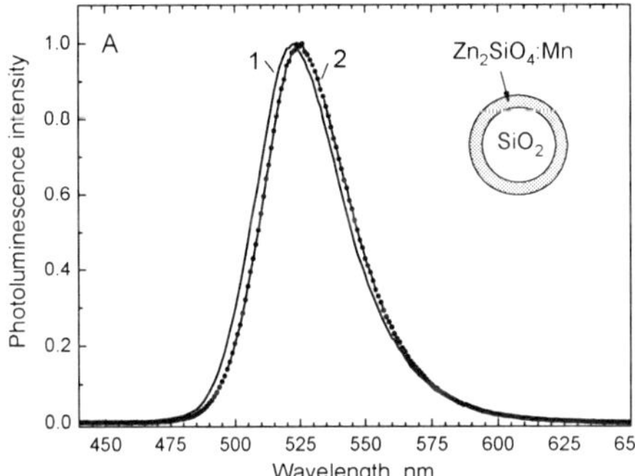

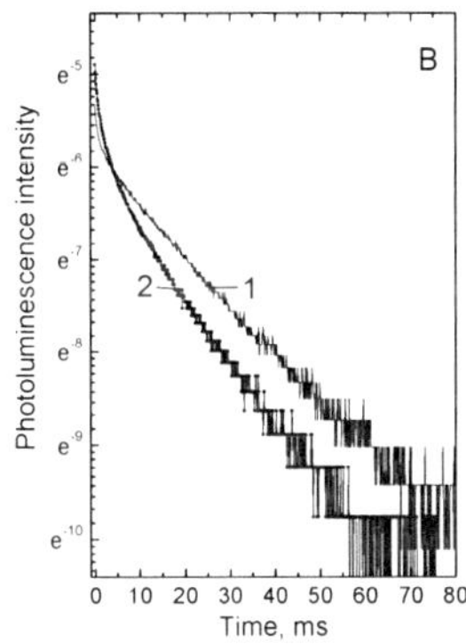

Figure 2. Photoluminescence (A) and photoluminescence decay (B) of two SiO$_2$/Zn$_2$SiO$_4$:Mn samples, annealed at 1100 °C, excited at 260 nm. Core diameter 197 nm. Sample 1 – doped with 0.9 mol% Mn, estimated shell thickness 5 nm, sample 2 - doped with 2.7 mol% Mn, estimated shell 10 nm.

(260 nm) the samples show the typical green luminescence of Mn^{2+} ions with the peak wavelength of 525 nm corresponding to the $^4T_1 - ^6A_1$ transition [11]. The intensity of this emission increases with the annealing temperature from 900 to 1100 °C. The maximum of the emission shows a slight red shift with growing Mn concentrations. The FWHM of the lines in fig 2A, measured in the samples doped with 0.9 and 2.7 mol% Mn (determined by the ICP analysis), is 40 nm. The decay curves (fig. 2B) are nonexponential at these concentrations, showing quenching of the emission due to the Mn-Mn interaction. The latter is relatively strong as the closest metal ion distances are 0.3 nm [12]. The decay of the forbidden d-d emission in single Mn^{2+} ions is expected to be simple exponential with a decay time of about 15 ms [13].

Practically all the samples exhibit some phosphorescent afterglow and are photostimulable with red light, if charged with UV light at 260 nm (corresponding to the charge transfer transition), or at higher photon energies (band-band transition). During charging, the electrons are captured by different traps, and may recombine with the created $(Mn^{2+} + h)$ pair later on. During charging with above-bandgap photon energies, the holes are probably captured by the Mn^{2+} ions. Figure 3 shows a typical phosphorescence decay curve of the core/shell

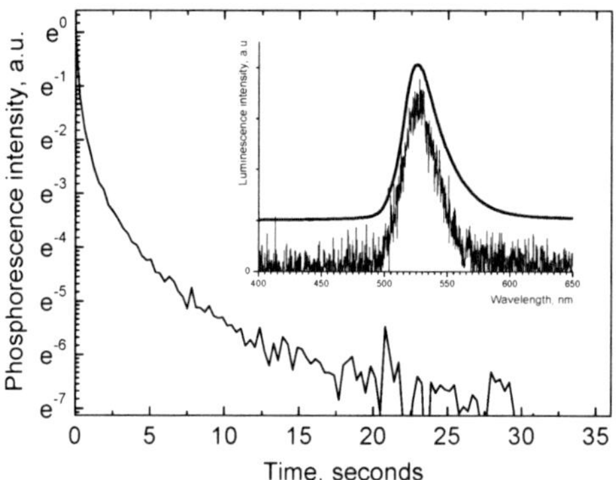

Figure 3. Afterglow decay of $SiO_2/Zn_2SiO_4:Mn^{2+}$ particles with the core of 110 nm, and shell thickness 10 nm. The detection at 520 nm began 20 ms after charging with 260 nm light. In the inset, the spectra of the afterglow and photoluminescence are compared (vertically shifted). T= 10 K.

particles, which is close to the t^{-1} law, characteristic to systems with a distribution of different traps. On the other hand, the similar decay law is obtained for the case of tunneling phosphorescence, which is probably the case in $Zn_2SiO_4:Mn^{2+}$ [14]. The fact that phosphorescence may be observed at low temperatures (see Figure 3, inset) speaks for the favour of tunneling as well. The inset in figure 3 shows the photoluminescence spectra and phosphorescence spectra of the Mn ions. The close similarity of the spectra gives evidence of the regular Mn ions taking part in the phosphorescence.

If charged with the 260 nm light or with the light of higher photon energy, the samples can be stimulated with red light to produce the green Mn^{2+} luminescence. This can be done after disappearing of the phosphorescence, even tens of minutes after the charging, meaning that the traps responsible for the PSL are still populated, when the phosphorescence traps are emptied. Another detail which speaks for the presence of different types of electron traps, is that in samples with different preparation conditions the ratio of phosphorescence and photostimulated luminescence intensity may be very different. The initial risetime of the photostimulated lifetime is correlated with the phosphorescence intensity: in samples with strong phosphorescence, the PSL signal rises slowly

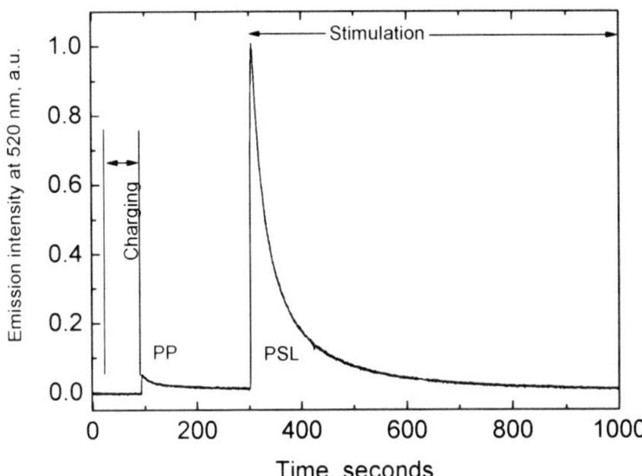

Figure 4. Afterglow (PP) and photostimulated luminescence (PSL) of the $SiO_2/Zn_2SiO_4:Mn^{2+}$ particles with core diameter of 197 nm. Mn doping concentration 0.9 mol%. Charging wavelength 260 nm, stimulation at 650 nm, detection at 520 nm. T= 300 K.

whereas the rise of the PSL in samples with low phosphorescence intensity is rapid. This shows that the stimulation of the PSL traps leads to a filling of the phosphorescence traps, competing with the recombination at the Mn ions.

CONCLUSIONS

Dispersable core-shell type particles of $SiO_2/Zn_2SiO_4:Mn$ with well-defined morphology and diameter under 300 nm were prepared by a Pechini-type sol-gel process. The zinc silicate shell has α-willemite structure, is highly luminescent, and enables separation and trapping of charges, if excited by the UV light. This leads to phosphorescence and, under red-light excitation, to photostimulated luminescence.

ACKNOWLEDGMENTS

This work was financially supported by the Fraunhofer-Gesellschaft zur Förderung der angewandten Forschung e.V., Munich, Germany (Grant No. 692034 and 663662), the Deutsche Forschungsgemeinschaft DFG (Grant No. GE 2118/3-1 and BA 2245/3-1), and the DESY, Hamburg, Germany (Project II – 20090087).

REFERENCES

1. J. Ziegler, S. Xu, E. Kucur, F. Meister, M. Batentschuk, F. Gindele, Th. Nann, *Adv. Mater.* **20**, 4068 (2008).
2. H. S. Mader, O. S. Wolfbeis, *Anal. Chem.* **82** (12), 5002 (2010).
3. H. Bannai, S. Levi, C. Schweizer, M. Dahan and A. Triller. *Nature Protocols.* **1**, 2628 (2006).
4. Y. I. Park, J. H. Kim, K. T. Lee, K.-S. Jeon, H. B. Na, J. H. Yu, H. M. Kim, N. Lee, S. H. Choi, S.-I. Baik, H. Kim, S. P. Park, B.-J. Park, Y. W. Kim, S. H. Lee, S.-Y. Yoon, I. C. Song, W. K. Moon, Y. D. Suh, T. Hyeon, *Adv. Mater.* **21**, 4414 (2009).
5. Q. le Masne, C. Chanéac, J. Seguin, F. Pellé, S. Maítrejean, J-P. Jolivet, D. Gourier, M. Bessodes. PNAS **104**, 9266 (2007).
6. Ph. Avouris, I. F. Chang, D. Dove, T. N. Morgan, Y. Thefaine, *J. Electronic Mat.* **10**, 887 (1981).
7. R. Selomulya, S. Ski, K. Pita, C. H. Kam, Q. Y. Zhang, S. Buddhudu, *Mater. Sci. Eng.* **B100**, 136 (2003).
8. R. Chen, Y. Wang, Y. Hu, Z. Hu, C. Liu, J. Lumin. **128**, 1180 (2008).
9. W. Stoeber, A. Fink and E. Bohn, *J. Colloid Interface Sci.* **26**, 62 (1968).
10. S. Dembski, S. Rupp, C. Gellermann, M. Batentschuk, A. Osvet, A. Winnacker, *J. Colloid Interface Sci.* **286**, 32 (2011).
11. P. Thioulouse, I. F. Chang, E. A. Giess. *J. Electrochem. Soc.* **130**, 2065 (1983).
12. T. C. Brunold, H. U. Güdel, E. Cavalli, *Chem. Phys. Lett.* **252** 112 (1996).
13. C. Barthou, J. Benoit, P. Benalloul, A. Morell, *J. Electrochem. Soc.* **141**, 524 (1994).
14. P. Avouris, T. N. Morgan, *J. Chem. Phys.* **75**, 4347 (1981).

Mater. Res. Soc. Symp. Proc. Vol. 1342 © 2011 Materials Research Society
DOI: 10.1557/opl.2011.997

Imaging Upconversion from NaYF$_4$:Er:Yb Nanoparticles on Au and Ag Nanostructured Substrates

Lanlan Zhong[1], QuocAnh Luu[2], Hari P. Paudel[3], Khadijeh Bayat[3], Mahdi Farrokh Baroughi[3], P. Stanley May[2] and Steve Smith[1,*]

[1]Nanoscience and Nanoengineering, South Dakota School of Mines and Technology
[2]Chemistry, University of South Dakota
[3]Electrical Engineering and Computer Science, South Dakota State University

*Corresponding Author: Steve_Smith@mailaps.org

ABSTRACT

Near-infrared-to-visible upconversion materials have many promising applications, including use in luminescent solar concentrators, in next-generation displays, and as biological labels. NaYF$_4$ nano-particles doped with Yb and Er exhibit efficient upconversion and are easily deployed in these applications. It is known that a rough metal surface may increase the yield of fluorescence of a nearby fluorophore, by local field enhancement due to plasmonic resonances, and by modification of the radiative rate(s) of the fluorophore. Thus, properly chosen metallic nanostructures can potentially increase the upconversion efficiency of lanthanide-doped nanoparticles, yet the optimal design of these nanostructures is still an active area of research. In our experiments, we use a spectroscopic imaging system to study the upconversion efficiency of NaYF$_4$: Er^{3+}/ Yb^{3+} through spatially-resolved upconversion spectra, using a custom-built scanning confocal microscope system with infra-red excitation, and wide-field fluorescence imaging. We present spectrally-resolved upconversion images of NaYF$_4$:Yb^{3+}/Er^{3+} nanoparticles on plasmonic substrates, including silver nanowires and patterned substrates of gold and silver, which show localized regions (~ 1μm) of relatively stronger intensity and modified upconversion spectra, and compare these to wide-field fluorescence images of samples with and without plasmonic substrates.

INTRODUCTION

Two photon processes could be very important in implementing third generation solar cell technologies, and can play an important role in characterizing the materials and devices necessary to implement them. Complementary approaches are spectral upconversion [1], intermediate band absorption [2,3], and multi-exciton generation [4,5]. These methods all increase the fraction of the solar spectrum which is converted to electricity, and / or increase conversion efficiency by utilizing / recapturing carrier excess energy normally lost to heat in the photovoltaic process. The present study concentrates on spectral upconversion, using NaYF$_4$ nanocrystals activated with Yb^{3+} and Er^{3+}. NIR-to-visible upconverters such as Yb,Er co-doped in NaYF$_4$ nanocrystals have potential as spectral converters and luminescent concentrators for solar cells [6], as well as other uses, such as nanolabels for biosensing or 3D displays [7].

However, a primary obstacle to the incorporation of upconversion phosphors into real devices has been the inability to obtain high upconversion efficiencies under modest excitation flux. It is now well-known that luminescence efficiencies of both organic and inorganic phosphors can be enhanced by their suitable proximity to a metal surface, due partly to an increase in the intrinsic radiative decay rates of the phosphor [7,8]. Previously, we used spectrally-resolved confocal microscopy to characterize the upconversion from Yb,Er: $NaYF_4$ nanocrystals on gold and silver nanostructured substrates [9]. In this work, we examine alternative nano-structured metallic substrates, including Ag nanorods, and 2D patterned periodic arrays of Au nano-pillars, fabricated by Electron Beam Lithography.

EXPERIMENT

$NaYF_4$:Er:Yb nanocrystals exhibiting strong infra-red to visible upconversion (UC NPs) were synthesized at USD, a typical TEM micrograph of these nanoparticles is shown in figure 1A. Details of the synthesis, physical and optical properties of these particles can be found in reference [9]. In the present work, we use a custom-built scanning confocal microscopy system to examine thin films of UC NPs embedded in PMMA, and assess possible enhancement in the upconversion intensity as a consequence of interactions with Ag and Au metallic nanostructures.

The loading of upconverting $NaYF_4$:$Yb^{3+}Er^{3+}$ nanoparticles in a polymer matrix results in a random distribution in UC-NPs, and, as such, significant variation in the density of UC NPs on the micron to sub-micron length scales (see figure 1). Using a custom-built, spectrally-resolved scanning-confocal microscope, we imaged the resulting spatial variations in upconversion intensity and spectra. The system consists of a closed-loop piezoelectric stage and high numerical aperture (Olympus UPLFLN-60X 0.9 NA) optics, epi-illuminating the sample with 980nm light from a diode laser, and collecting the subsequent upconverted light in shared-aperture mode. As seen in figure 1A, variations in UC NP density are superimposed on the nano-structured metallic substrates. Thus, any observed enhancement in the UC in a given region may have its origin in the proximity of the UC-NPs to local enhancement in the electromagnetic field

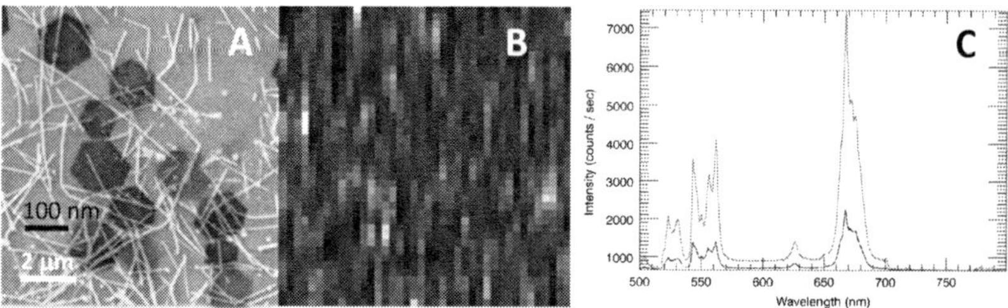

Figure 1: A: SEM micrograph of ~ 20nm diameter nanowires superimposed on TEM micrograph of ~ 90-100nm $NaYF_4$:Er:Yb nanocrystals (scale bars: white → SEM, black → TEM), B: spatially-resolved upconversion luminescence at peak wavelength in emission spectra near 670nm, C: representative fluctuations in upconversion spectra taken from dark and bright regions of image shown in B.

due to plasmon resonances in the metallic substrate, modifications in radiative rates associated with the upconversion process due to the metal, and / or local fluctuations in UC-NP density. Spectroscopic imaging provides a massively parallel means of assessing both the average UC intensity enhancement and fluctuations due to these variations. These fluctuations are clearly seen in the image of Fig 1B and the spectra of select points in this image, shown in Figure 1C.

RESULTS AND DISCUSSION

Neglecting spatial correlations necessarily contained in the images, a statistical analysis of the images obtained in these experiments can give information regarding the average and extremal values of NIR to visible upconversion enhancement. We use a statistical analysis of the images obtained in our experiments to assess these variations. We compared similarly prepared samples consisting of ~ 100nm PMMA films loaded with $NaYF_4:Er:Yb$ nano-particles, described in detail in reference [9], which were spin-coated on glass substrates. A control sample, consisting of only an upconverting layer, and a similar sample coated with a layer of silver nanowires (Ag NWs) were imaged as in figure 1, and the distribution of intensities of upconverted light obtained everywhere in the images are shown below in figure 2.

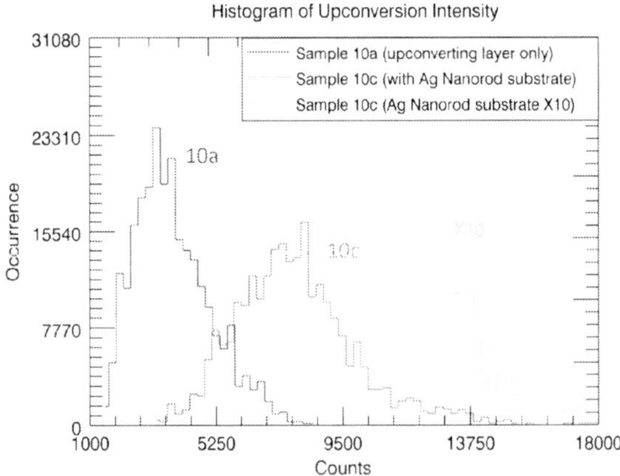

Figure 2: Histograms of upconversion intensity images formed from thin layers of UC NPs embedded in PMMA and spin-coated on glass without Ag nano-wire substrate (sample 10a), and with Ag nano-wire substrate (sample 10c).

In examining these distributions, one notes the well-defined averages, indicating the most-probable enhancement factor of about 2.5X. Noting the asymmetry in the red and blue distributions, one can conclude that the optimal density of UC NPs and Ag NWs has not been achieved. That is why the tail of the red curve extends far to the right towards higher upconversion intensities. This tail is magnified by 10X for clarity. Evidently, there are particular arrangements of UC NPs and neighboring Ag NWs which yield much higher intensity enhancement, but these are less probable. The results motivate a search for the optimal surface densities / areal coverage of UC NPs and/or Ag NWs.

As a first step towards this goal, we held the UC NP density constant and prepared samples with varying surface density / areal coverage of Ag NWs. A comparison of the statistical analysis of the spectroscopic images obtained for these samples, similar to the image shown in figure 1B, are shown below in figure 3. It can be seen by comparison of these distributions that higher surface coverage of Ag NWs results in larger upconversion intensities. Also, we note the distribution at 4X density appears symmetric, which implies that we may be close to optimal density of Ag NWs for the given UC NP density. Obviously, further optimization of both densities would be required to find the optimal sample parameters. However, these results suggest that spectroscopic imaging and statistical analysis may provide an assay of these possible arrangements, and is a useful tool in optimizing such UC NP / Ag NW systems.

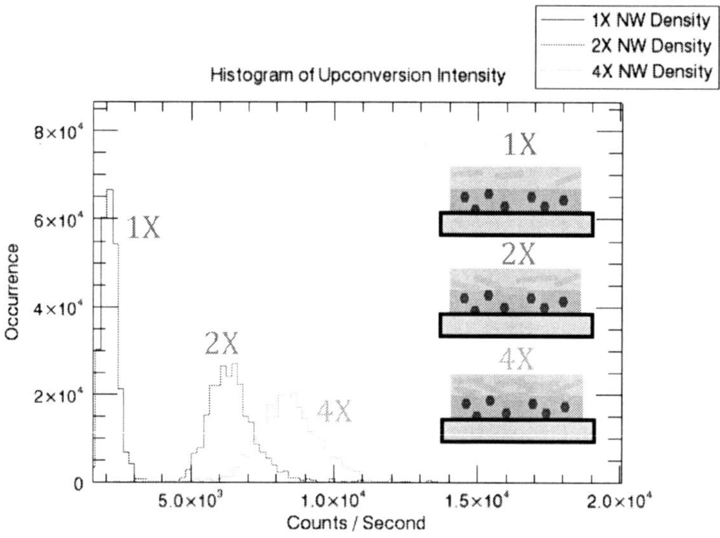

Figure 3: Statistical analysis of spectroscopic images for 3 UC NPs embedded in PMMA samples where coated with varying density of Ag NWs, by factors of 1X, 2X and 4X.

We used our imaging methods to assay the influence of another type of metallic nanostructured substrate on the efficiency of the upconversion intensity from UC NP films. Specifically, nano-patterned Au nano-pillar arrays, fabricated by electron beam lithography. Films of UC NPs embedded in PMMA were deposited on these substrates by spin-coating, similar to the preparation of the UC NP / Ag NW samples. Figure 4 shows an image formed by scanning across the edge of a patterned array of Au nanopillars, the scan window arranged such that half the image is formed in an area containing the nano-pillared sample, and half the image is formed over bare glass, as shown in figure 4A&B, where the entire sample was uniformly coated with UC-NPs suspended in a polymer matrix. Figure 4C shows typical spectra taken from the two regions. By inspection of figure 4B, we see that the intensity is relatively uniform in the area containing the nano-pillars, and similarly for the glass-only portion of the scan window. Comparing the spectra in figure 4C, we obtain a ratio of approximately 2.5. Given the uniformity in each region, the enhancement factor(s) obtained should be representative of the sample.

In figure 4D&E, we show a white-light image and spectral scan image formed within the patterned portion of the sample. Here, again, the intensity is rather uniform, except for the localized "defect" observed in the lower left portion of the image, which, by comparing the spectra in figure 4F, is enhanced approximately 12 times. Similar localized enhancement was observed in other areas of these samples, and comparison of these regions with optical micrographs suggests the origin of the localized enhancement may be a "missing pillar". It is well-known from the study of two-dimensional photonic crystals that such a missing lattice point

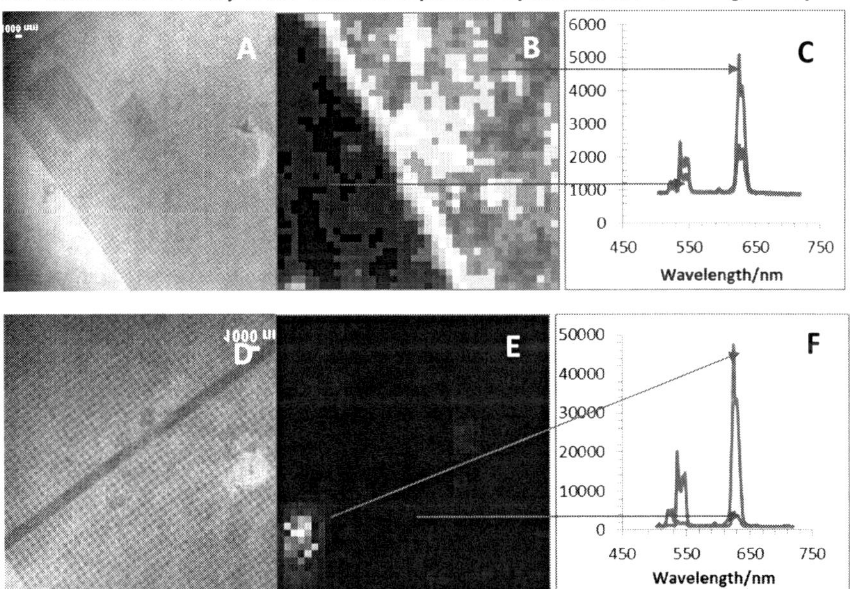

Figure 4: (a): White-light image of Au nano-pillar sample, (b): UC image which traverses the boundary of the patterned portion of the sample, (c): representative spectra of patterned and unpatterned portions of the sample, (d): White-light image of Au nano-pillar sample showing possible defect, (e): UC image of patterned portion of the sample containing a localized "defect", (f): representative spectra of "defect" area.

can form a local resonance, and similar effects may be possible with respect to our periodic nano-pillar substrates, however, we have not yet investigated these effects in detail.

CONCLUSIONS

We demonstrated spectroscopic imaging and statistical analysis of upconversion intensity maps obtained from UC NPs embedded in PMMA films, supported on Au and Ag metallic nano-structured substrates, can be an effective assay to characterize the achievable enhancement in infra-red to visible upconversion. In particular, we saw that random Ag nanowire substrates can achieve an average of 2.5X enhancement in the upconversion from UC NP / PMMA films, and up to 6X enhancement at specific points in the films. Statistical analysis can give a good measure of the range and average of the enhancement achieved, while the shape of the distributions can guide sample design, as demonstrated by varying the density of Ag nanowire coverage and comparing the distributions of upconversion intensity obtained from spectroscopic images. We also analyzed Au periodic nano-pillar arrays, and observed a local defect effect, which can result in up to 12X enhancement at specific sites in these arrays, the origin of which was not revealed in these experiments.

ACKNOWLEDGMENTS

Funding provided by NASA award #NNX09AP67A, NSF award #'s 0619890 (DMR), 0903804 (EPSCoR), DOE through NREL subcontract # ZCO-7-77379-01 and the State of South Dakota. The gold nano-pillared arrays were provided by M. Baroughi of South Dakota State University.

REFERENCES

[1] S. Baluschev, T. Miteva, V. Yakutkin, et. al., "Up-Conversion Fluorescence: Noncoherent Excitation by Sunlight," *Phys. Rev. Lett.* **97** 143903 (2006).
[2] Antonio Luque and Antonio Marti', "Increasing the Efficiency of Ideal Solar Cells by Photon Induced Transitions at Intermediate Levels," *Phys. Rev. Lett.* **78** (26) 5014 (1997).
[3] R.W. Peng, M. Mazzer, K.W.J. Barnham, "Efficiency enhancement of ideal photovoltaic solar cells by photonic excitatons in multi-intermediate band structures," *Appl. Phys. Lett.* **83** (4) 770 (2003).
[4] R. D. Schaller and V. I. Klimov, "High Efficiency Carrier Multiplication in PbSe Nanocrystals: Implications for Solar Energy Conversion," *Phys. Rev. Lett.* **92** 186601 (2004).
[5] Richard D. Schaller and Victor I. Klimov, "Non-Poissonian Exciton Populations in Semiconductor Nanocrystals via Carrier Multiplication," *Phys. Rev. Lett.* **96** 097402 (2006).
[6] G.B. Smith," Materials and systems for efficient ... delivery of daylight" *Solar Cell Mat.*, **84** 395 (2004).
[7] Lakowicz, J.R , "Radiative decay engineering: biophysical...applications" *Analytical Biochemistry* **298** 1 (2001).
[8] K. Aslan, J,R. Lakowicz, C.D. Geddes, "Metal-enhanced fluorescence using anisotropic silver nanostructures: critical progress to date", *Anal. Bioanal. Chem.* **382** 926 (2005).
[9] Cuikun Lin, Mary T. Berry, Robert Anderson, Steve Smith, and P. Stanley May, "Highly Luminescent NIR-to-Visible Upconversion Thin Films and Monoliths Requiring No High-Temperature Treatment," *Chem. of Materials* **21** (14), 3406-3413 (2009).

Rare Earth Ions in Group III - Nitrides

Mater. Res. Soc. Symp. Proc. Vol. 1342 © 2011 Materials Research Society
DOI: 10.1557/opl.2011.998

Ultraviolet Light Emitting Devices Using AlGdN

Takashi Kita[1], Shinya Kitayama[1], Tsuguo Ishihara[2], Hirokazu Izumi[2], Yoshitaka Chigi[3], Tetsuro Nishimoto[3], Hiroyuki Tanaka[3], and Mikihiro Kobayashi[3]
[1]Department of Electrical and Electronics Engineering, Graduate School of Engineering, Kobe University, Rokkodai 1-1, Nada, Kobe 657-8501, Japan
[2]Hyogo Prefectural Institute of Technology, Yukihira 3-1-12, Suma, Kobe 654-0037, Japan
[3]YUMEX INC., Itoda 400, Yumesaki, Himeji, Hyogo 671-2114, Japan

ABSTRACT

We developed ultra-violet field-emission devices using rare-earth nitrides of $Al_{1-x}Gd_xN$ grown by a reactive radio-frequency magnetron sputtering technique. The $Al_{1-x}Gd_xN$ phosphor film excited by high-energy electrons shows a resolution limited, narrow intra-orbital luminescence from Gd^{3+} ions at 318 nm. The devise characteristics depend on injected current and acceleration voltage, which were analyzed by considering multiple excitation process of injected high-energy electrons.

INTRODUCTION

Ultra-violet (UV) light is indispensable for various industrial processes such as lithography, curing, polymerization using photochemical reactions, sterilization, and medical treatment. For these applications, many kinds of discharge lamps using mercury have been widely used in such fields. However, mercury is recognized as a chemical of global concern due to its long-range transport in the atmosphere, its persistence in the environment, its ability to bio-accumulate in ecosystems and its significant negative effect on human health and the environment. For human life free from mercury, it is strongly desired to develop novel UV lighting source which can replace currently available mercury lamps. Recently, nitride-semiconductor based UV light emitting diodes (LEDs) are attracting strong interest.[1] In LEDs, the emission wavelength can be controlled by the optical band gap, and efficient light emission is realized by utilizing quantum structures. However, the spectral line width limited by the thermal carrier distribution near the band edge causes a relatively broad emission-band width being generally over 10 nm. Moreover, when we need to irradiate large area with the UV light, the flux from a small LED chip must be controlled to prepare uniform power distribution on the large area. Such point source of LEDs with the broad band emission is unfortunately limiting the application. For photo lithography to obtain the high resolution and medical applications to avoid unexpected side effects, light panel with a narrower spectral width is required.

Instead of using band-edge emission, we focus on the use of narrow band emissions from the intra-orbital electron transitions of rare earth ions.[2-4] Rare earth ions are widely used in various phosphors, solid state laser crystals, and optical amplifiers. In this work, we focus on Gd which emits the luminescence near 300 nm. To avoid absorption of the UV emission, the host material must have a wide bandgap more than 4 eV. AlN is suitable for this requirement.[5-8] The band gap is approximately 6.2 eV at room temperature, and Gd can replace Al. The thermal conductivity of AlN is quite high, which is comparable with metals. Also, AlN is a thermally and chemically stable material, and, of course, Gd compound are not harmful. Generally, Gd doping has been carried out by using ion-implantation and sputtering growth techniques. In contrast to the ion-implantation, the sputtering growth is expected to realize high quality crystals without annealing. In this work, we fabricated $Al_{1-x}Gd_xN$ thin films using an ultra-pure sputtering growth technique and developed field-emission devices. We discuss detailed device characteristics and the excitation process.

GROWTH OF $Al_{1-x}Gd_xN$ THIN FILMS

We fabricated $Al_{1-x}Gd_xN$ thin films using the reactive magnetron sputtering technique performed under molecular-beam epitaxial quality.[2-4] A growth chamber equipped with three targets was separated from a substrate-introduction chamber. The base pressure of the growth chamber was less than approximately 5×10^{-6} Pa. We used a 6N mixed gas of argon and nitrogen for the reactive growth. The metal targets used here were 4N Al and 3N Gd. When growing $Al_{1-x}Gd_xN$ alloy, we set Gd metal tips on the Al target. The Gd concentration was controlled by changing the number of tips. All samples were grown on fused silica substrates. The growth temperature was $200°C$. First we grow a 800-nm AlN buffer layer, and, then, AlGdN film was deposited.

The deposited thin film is a c-axis oriented polycrystal.[2] The x-ray rocking curve of the c-plane diffraction indicates that the tilting angle of the c axis was within 6 degrees. In this work, we grew $Al_{1-x}Gd_xN$ alloy films with various GdN-mole fractions. With increasing the GdN-mole fraction, the diffraction peaks shift toward the low angle side, because the large Gd atoms expands the lattice constant of AlN. The radial structure function of $Al_{1-x}Gd_xN$ examined by extended X-ray absorption fine structure (EXAFS) measurements is clearly different from that of GdN. Thus, GdN clusters were not confirmed in $Al_{1-x}Gd_xN$ even for a high mole fraction x of 6%.

OPTICAL PROPERTIES OF $Al_{1-x}Gd_xN$ THIN FILM

The band edge structure of $Al_{1-x}Gd_xN$ was investigated by using photoluminescence-excitation (PLE) measurements at room temperature. The excitation was carried out using the monochromatic light of a Xe lamp. The wavelength resolution in the PLE measurements was approximately 2.5 nm. A typical PL spectrum of $Al_{0.999}Gd_{0.001}N$ excited at 200 nm is displayed in the inset of Fig. 1. The sharp peak appeared at 318 nm, which corresponds to the transition from the lowest excited state $^6P_{7/2}$ to the ground state $^8S_{7/2}$. The PLE spectra detected at 318 nm for $Al_{0.999}Gd_{0.001}N$ (solid line) and $Al_{0.99}Gd_{0.01}N$ (dashed line) are shown in Fig. 1. The strong absorption near 200 nm corresponds to the band edge of $Al_{1-x}Gd_xN$. With the increase in the GdN-mole fraction, the band gap was found to be remarkably reduced; only 1 mol% GdN causes the large shift more than 5 nm. It is noted that the band gap of GdN at room temperature is in the infrared region.[9] Thus, the band gap of $Al_{1-x}Gd_xN$ is drastically reduced by containing a small amount of Gd.

We confirmed weak absorption signals below the band edge as shown in Fig. 2. These are attributed to the excited states of Gd^{3+} ion.[6] Such direct excitation components become significant in the high Gd-concentration film. According to this excitation spectrum, the indirect excitation process is considered to be dominant in our system.

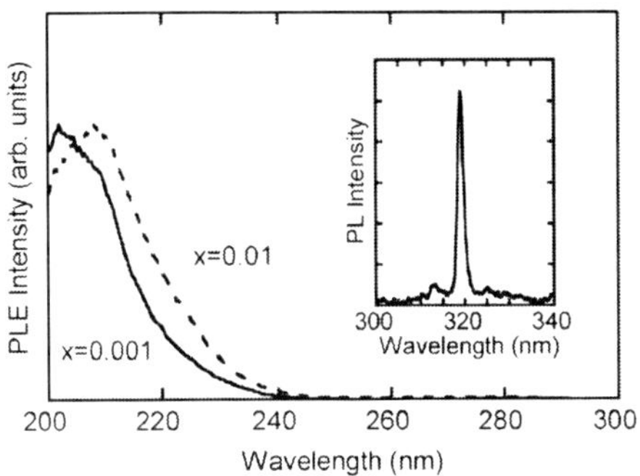

Figure 1. PL and PLE spectra for $Al_{0.999}Gd_{0.001}N$ (solid line) and $Al_{0.99}Gd_{0.01}N$ (dashed line). The inset shows a typical PL spectrum of $Al_{0.999}Gd_{0.001}N$ excited at 200 nm.

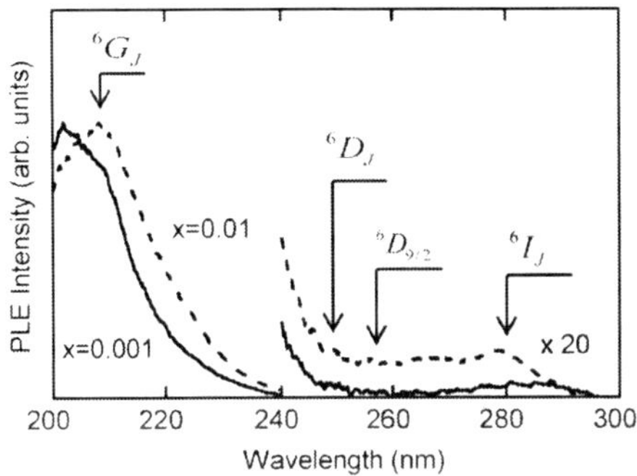

Figure 2. Direct excitation signals appeared below the band edge.

FIELD-EMISSION DEVICE CHARACTERISTICS

In this work, we developed field emission devices using $Al_{1-x}Gd_xN$ thin films. This device consists of an electron emitter, insulating spacers, electrodes of cathode, grid, and anode, and $Al_{1-x}Gd_xN$ phosphor thin film. Here, we used carbon nanofiber/elastomer composite as the electron emitter.[10] The emission current was controlled by the grid-gate voltage. This device was operated at room temperature in vacuum of 6×10^{-4} Pa. Figure 3 is a typical luminescence spectrum obtained at 5 kV and 0.1 mA. The spectrum shows a strong, sharp peak near 320 nm. The high resolution spectrum is shown in the inset. The strong peak was observed at 318 nm. The line width is approximately 1 nm. This peak consists of several lines caused by the Stark splitting. A weak signal appeared at 312 nm can be attributed to the transition from the second lowest excited states $^6P_{5/2}$ to the ground state $^8S_{7/2}$.

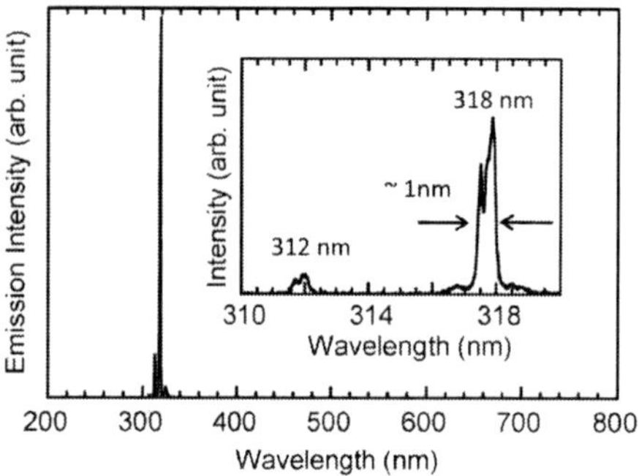

Figure 3. Typical luminescence spectrum obtained at 5 kV and 0.1 mA.

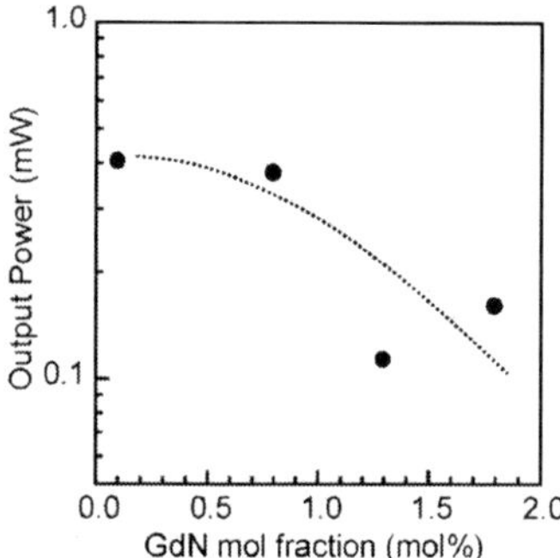

Figure 4. GdN-mole fraction dependence of the output power.

Figure 4 shows the GdN-mole fraction dependence of the emission intensity. With increasing the mole fraction, the intensity was decreased. This is considered to be due to so called "concentration quenching". In the following detailed characteristics, we used $Al_{0.999}Gd_{0.001}N$.

The injection current dependence of the output power as a function of the acceleration voltage is shown in Fig. 5. The output power increases with the current and shows saturation behaviors. The saturation behavior depends on the excitation voltage. The maximum output power was approximately 0.5 mW/cm^2 and the power efficiency at this maximum output condition was approximately 0.01%. According to this saturation value, the density of the active Gd enters was estimated to be less than 1% of the doped Gd, which may support that the excitation mechanism is indirect; loss in the energy transfer from the host material to Gd center is significant. This saturation behavior can be analyzed by using a well known relation of the excitation cross section[5,11];

$$I \propto \frac{\sigma\tau(J/e)}{\sigma\tau(J/e)+1} \qquad (1)$$

Where σ is the excitation cross section and τ is the lifetime of the excited state. The evaluated cross sections are in the order of 10^{-13} cm^2. These values are very large as compared with previously reported data.[11] Furthermore, the excitation cross section was found to depend on the acceleration voltage. Therefore, we considered that effective current is considered to be increased by multiple excitation process of the injected excess high-energy electrons.

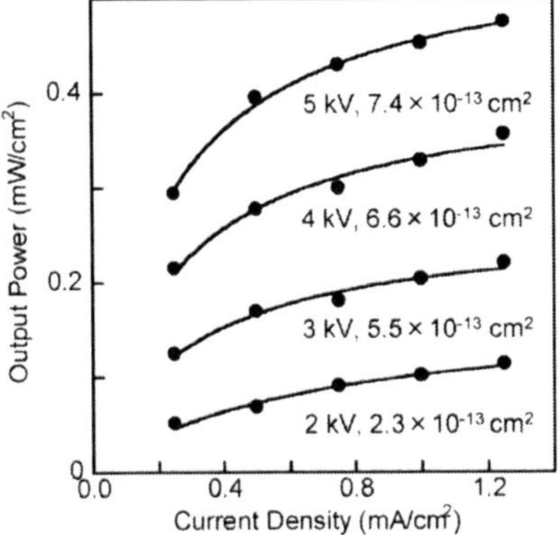

Figure 5. Injection current dependence of the output power as a function of the acceleration voltage. The excitation cross section was found to depend on the acceleration voltage.

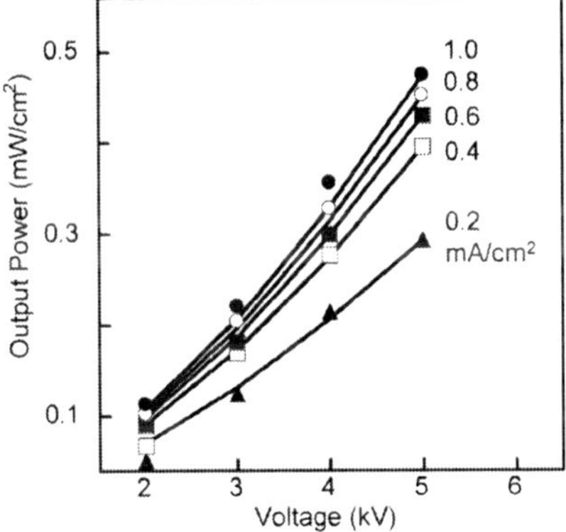

Figure 6. Acceleration voltage dependence of the output power.

The acceleration voltage dependence of the output power is displayed in Fig. 6. The output power increases with the acceleration voltage even at the same injection current. This is a direct evidence of the multiple excitation process. We analyzed these results according to the depth profile of injected electrons. Generally, high energy electrons can penetrate into the deep region in the film, and the absorbed energy density distribution depends on the acceleration voltage. If we assume a maximum absorbed energy density as inferred from the small saturation power, the excitation intensity is proportional to this area below the maximum absorption level where the absorbed energy can transfer to Gd ions. The calculated relative intensities based on this model are indicated by solid lines. These results agree well with the experimental data except for the data obtained at the low acceleration voltage. Deviation observed at the low acceleration voltage is considered to be due to energy loss caused near the surface of the phosphor film because of the shallow penetration.

CONCLUSIONS

We demonstrated mercury-free narrow-band ultraviolet emission using polycrystalline $Al_{1-x}Gd_xN$ phosphor thin film grown by using the reactive magnetron sputtering technique. The resolution limited, narrow intra-orbital luminescence from Gd^{3+} ions was observed at 318 nm. Developed field-emission devices were systematically investigated, and the devise characteristics depending on the injected current and acceleration voltage were analyzed considering multiple excitation process of the injected high-energy electrons.

REFERENCES

1. H. Hirayama, S. Fujikawa, N. Noguchi, J. Norimatsu, T. Takano, K. Tsubaki, N. Kamata, *Phys. Status Solidi* (a) **206**, 1176 (2009).
2. T. Kita, S. Kitayama, M. Kawamura, O. Wada, Y. Chigi, Y. Kasai, T. Nishimoto, H. Tanaka, and M. Kobayashi, Appl. Phys. Lett. **93**, 211901 (2008).
3. S. Kitayama, T. Kita, M. Kawamura, O. Wada, Y. Chigi, Y. Kasai, T. Nishimoto, H. Tanaka, and M. Kobayashi, IOP Conf. Series: Materials Science and Engineering **1**, 012001 (2009).
4. T. Kita, S. Kitayama, H. Yoshitomi, T. Ishihara, H. Izumi, Y. Chigi, Y. Kasai, T. Nishimoto, H. Tanaka, M. Kobayashi, J. Ceramic Processing Research **12**, s73 (2011).
5. U. Vetter, J. Zenneck, and H. Hofsäss, Appl. Phys. Lett. 83 (2003) 2145.
6. J. B. Gruber, U. Vetter, H. Hofsäss, B. Zandi, M. F. Reid, Phys. Rev. B **69** 195202 (2004).
7. J. M. Zavada, N. Nepal, J. Y. Lin, H. X. Jiang, E. Brown, U. Hömmerich, J. Hite, G. T. Thaler, C. R. Abernathy, and S. J. Pearton, Appl. Phys. Lett. **89**, 152107 (2006).
8. M. Maqbool, I. Ahmad, H. H. Richardson, and M. E. Kordesch, Appl. Phys. Lett. **91**, 193511 (2007).
9. H. Yoshitomi, S. Kitayama, T. Kita, and O. Wada, Phys. Rev. B, **83**, 155202 (2011).
10. N. Kishi, T. Kita, A. Magario, and T. Noguchi, J. Appl. Phys. **109**, 074307 (2011).
11. A. Koizumi, Y. Fujiwara, A. Urakami, K. Inoue, T. Yoshikane, and Y. Takeda, Appl. Phys. Lett. **83**, 4521 (2003).

Mater. Res. Soc. Symp. Proc. Vol. 1342 © 2011 Materials Research Society
DOI: 10.1557/opl.2011.1050

Theoretical investigation of Er-O co-doping in hexagonal GaN

Simone Sanna, Uwe Gerstmann and Wolf Gero Schmidt
Lehrstuhl für Theoretische Physik, Universität Paderborn, Warburger Straße 100, 33098
Germany.

ABSTRACT

The co-doping of hexagonal GaN with Er and O is investigated by means of density
functional calculations. Predominantly Er-O defect-pairs characterized by a binding energy
around 0.5 eV are formed. Different geometric configurations with various orientations (i.e. axial
and basal pairs with C_{3v} or C_{1h} symmetry) are expected with similar formation energies.
Independent of the particular configuration, the presence of oxygen does not deeply affect the
atomic structure and the electronic charge distribution around the Er centers. The relatively high
binding energy suggests that Er-O pairs should survive thermal treatment. An investigation of the
binding energy per bond indicates that on the other hand Er-O_x complexes (x=2,3,4) are not
likely to be formed (differently from Er-O co-doped Si). Rather, as long as the oxygen fluence
does not overtake the Er fluence, different oxygen ions will be bound to different Er-centers.

INTRODUCTION

Rare earth (RE) doped semiconductors are attracting an increasing interest due to their
application in light emitting devices and lasers. The optical emission from RE ions consists of
very sharp lines (ranging from the UV to IR) which are not affected by the host material and
whose wavelength is determined by the energy of the corresponding transition between the *4f*-
states [1]. Different factors, such as the thermal quenching of the luminescence, the energy
transfer from the host to the rare-earth *4f*-shells and the efficiency of the luminescent transitions
(i.e. the radiative/non-radiative decay ratio) have represented up to date important limits to the
realization of customer-ready commercial devices [2]. However, several steps towards the
understanding of the excitation and emission mechanisms in different hosts as well as towards
the identification of the luminescent centers have been done recently [2,3]. Concerning the
optimization of the luminescent devices, both the use of high quality RE-doped samples and co-
doping with light elements such as C and O have lead to important improvements [5]. It has been
demonstrated that in semiconductors such as Si and GaAs the co-implantation of rare earths with
oxygen is beneficial for the luminescence [4-7]. In Si and GaAs the co-doping has the dual
purpose to increase the luminescence intensity and to reduce the temperature quenching of the
emission. This is achieved by the formation of particular RE-O complexes, e.g. Er-O_2 [17,18] in
GaAs and Er_2O_3 in Si [19,20]. The effect of co-doping in other hosts with higher electronic band
gap is still under debate, though. The knowledge of the effect of oxygen co-doping in RE doped
GaN and AlN would be highly desirable, as these semiconductors are particularly suitable hosts
for the RE. Indeed, it has been shown that the quenching of the emission at room temperature
decreases with increasing band gap of the host material [8]. Unfortunately despite the effort of
different research groups no consistent picture could be achieved and contradictory reports are
available in the literature. Torvik *et al.* reported an increase of photoluminescence (PL) intensity

by 20 times upon co-doping Er implanted samples with oxygen at 10:1 (O:Er) ratio [9]. They proposed different mechanisms to explain the beneficial effect of oxygen, including the formation of Er_2O_3 structures [10]. Citrin *et al.* investigated Er-O codoped GaN-films by x-ray absorption and electroluminescence measurements. They found the luminescence properties as well as the local structure of the RE to be unaffected by the amount of O present in the host and used simple chemical concepts (such as ionic bond strength and ion size) to explain the striking differences between Er-O co-doped Si and GaN samples [11]. Alves and co-workers combined PL and Rutherford backscattering spectrometry (RBS) to investigate Er-doped and Er-O co-doped GaN samples. They found the luminescence intensity of co-doped GaN to be lower compared with samples without oxygen. This fact was explained by the formation of stable non-radiative Er-O complexes during annealing [12,13]. Finally, Zavada *et al.* found important differences in the PL of samples measured under below-gap and above-gap excitation. With below-gap excitation, the PL intensity of Er doped GaN with a strong O and C background was found two orders of magnitude larger than in samples with low O and C backgrounds. With above-gap excitation, the PL was greatly reduced with respect to the below-gap excitation, and was independent from the O and C concentration. The results for below-gap excitation were explained by the presence of mid-gap states, which provide efficient pathways to the Er^{3+} ions [14,15]. The only theoretical investigation of RE-O complexes in GaN is due to Filhol and co-workers was restricted to *pairs*, whereby the LDA-frozen core approach was used to model Eu, Er, and Tm doped GaN with and without oxygen. Independent of the RE, they found an oxygen-related donor level above the host conduction band and no mid-gap states [16]. Summarizing, the issue of RE-O co-doping in GaN seems not to be settled and much work on this field has still to be done. In this work we investigate the formation and microscopic structure of Er-O defect *complexes* in hexagonal GaN by first-principles calculations. Our calculations reveal the occurring defects, their symmetry and thermal stability, which are fundamental parameters in the luminescence process. On the basis of our model we discuss the defect complexes proposed to explain the luminescence intensity changes upon oxygen co-doping.

THEORY

Density functional theory (DFT) within the generalized gradient approximation (GGA) as implemented in the VASP simulation package [21] has been used to perform total energy calculations. Different defect configurations are modeled within supercells of hexagonal symmetry containing 256 atoms. A Γ-centered equidistant $4 \times 4 \times 4$ k-point mesh was used to carry out the integration in the Brillouin zone. The energy-cutoff for the plane wave basis was 400 eV. We used the PW91 formulation of the exchange-correlation functional [22] and projector augmented wave (PAW) potentials [23], whereby the Ga $3d$ and the Er $4f$ electrons were included in a pseudo-atomic core. The PAW for Er was created assuming the rare earth to exist in trivalent oxidation state only (Er^{3+}). In this configuration the $4f$-shell hosts 11 electrons (instead of 12 of the atomic configuration) and its occupation cannot be modified by the local environment of the RE or by chemical doping. Even if this approach (known as frozen-core LDA) might not be appropriate to describe the electronic structure for each value of the Fermi level within the GaN band-gap [24,25], it usually yields reliable geometries [16], which are the main goal of this work.

DISCUSSION

Electronic channeling (EC) studies have shown that co-implanting erbium together with oxygen or carbon does not result in the occupation of fundamentally different lattice sites of Er [9,26]. Differences between the co-implanted and the Er-implanted samples are within the error bars, which means that any difference in luminescence caused by co-implantation of Er with O or C into GaN cannot be attributed to a change in the lattice site of Er, in contrast to the case of Er in silicon [28]. On the other side, it is also known from previous studies that oxygen in GaN forms predominantly O_N substitutionals [29]. We simulate therefore structures with the rare earth atom on a Ga site and the oxygen substituting one or more of the neighboring N ligands.

Er-O pairs

Due to the hexagonal symmetry of the GaN host, there are two inequivalent places for the nitrogen site around the RE_{Ga} substitutional. This means that the RE-O complexes can occur in two different configurations, depending on the substituted nitrogen atoms. If the substituted atom is one of the three equivalent nitrogen (see Fig. 1a), the defect will only have a symmetry plane (which contains the Er-O bond and the c-axis) and therefore the C_{1h} symmetry (*basal* configuration). If the substituted nitrogen is the one along the crystal c-axis (*axial* configuration) the defect complex can have the C_{3v} (see Fig. 1b) or the C_{1h} symmetry, if Jahn-Teller distortion occurs. Whether all the minima are formed or not and their relative stability depend on the occupation of the gap states.

We find the basal pairs (Fig. 1a) to be energetically favored, but only slightly by 80 meV over the other configurations. This means that also defect complexes in the axial configurations have to be expected in the co-doped samples. This is particularly true if other factors besides the formation and binding energy influence the created defects, as discussed in the next section. Indeed, in the case of implanted samples, the direction of the impinging ions is known to favor basal pairs over axial pairs and vice versa [27]. The interatomic distances of the axial and basal pairs are reported in Fig. 1.

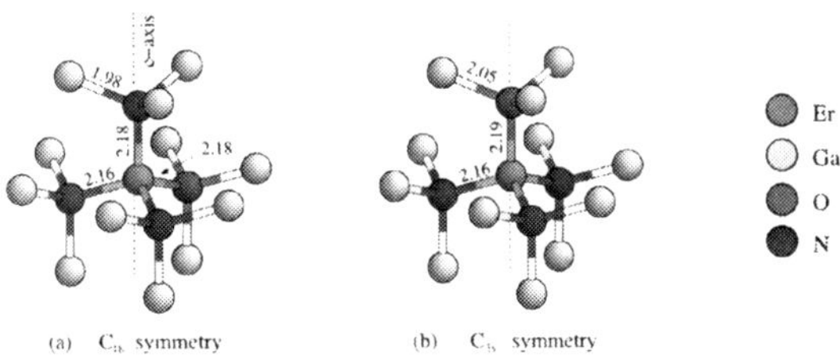

Figure 1. Er_{Ga} O_N defect centers in hexagonal GaN can assume two kind of configurations, basal (a) with symmetry C_{1h} and axial with symmetry C_{3v} (b) or C_{1h}, if Jahn-Teller distortion occurs. All lengths in Å.

We notice that in general the lattice distortion introduced by the co-doping is only marginal. The O-Ga bond lengths are only 3.5% bigger than the N-Ga bonds in the pristine host, and the Er-O bond lengths (2.18 Å and 2.19 Å for axial and basal pairs) do not really differ from the Er-N distance in isolated Er_{Ga} substitutionals (2.18 Å and 2.19 Å for axial and basal pairs). This is probably due to the similarity in the size (r_{cov}=0.73 and 0.75 Å) and electronegativity (3.5 and 3.0 in the Pauli scale) between oxygen and the substituted N. However, while in isolated Er_{Ga} substitutionals Er stays almost exactly at the Ga site, complexed with O it results displaced by 0.1 Å (basal pairs), outwards in the Er-O direction. Concerning the electronic charge transitions, we find Er-O to behave as isolated O_N substitutional, i.e. an ideal donor with a charge transition $\varepsilon(+/0)$ almost resonant with the GaN conduction band. An analysis of the total electronic charge reveals that the charge distribution on the N-ligands in isolated Er_{Ga} substitutionals and in Er-O complexes is substantially the same. This is a confirmation of the fact that the extra electron is not bound, but is rather localized. It does not take part to the bond and is a proper free carrier. We want to remind, however, that due to the frozen-core approach of this investigation, we are not able to model any possible charge transition involving the Er $4f$-states.

Er-O$_x$ complexes

In semiconductors such as Si (or GaAs) co-doping of Er with O is essential to the RE luminescence. In a Si-host that is O-rich, Er getters the O and forms soluble, optically active defects resembling erbium oxide. The amount of O in Si thus affects not only the optical activity of Er but the amount of Er that can be incorporated in Si by avoiding the formation of (optically inactive) Er silicide precipitates [11]. In order to investigate the possible formation of Er-O$_x$ complexes in GaN, we have modeled Er-centers surrounded by two, three (two configurations are possible in each case) and four oxygen ions. We observe that increasing the oxygen amount also increases the lattice distortion and the formation energy of the complexes. This occurs probably as Er and O try to reach a configuration similar to that in erbium oxide. As the formation energy of defect complexes involving a different number of substitutionals is not directly comparable, we use the binding energy as stability parameter. The Er-O binding energy decreases almost linearly with the number of oxygen atoms, so that once the Er-impurity has bound one oxygen, the energy gain in binding a second oxygen will be lower by around 100, 260 and 370 meV, respectively (see Fig 2). In other words, at moderate O doping concentrations, predominantly Er-O pairs are formed. Only in the regime of extremely high O concentration, in almost diluted material, a considerable amount of Er-O$_x$ complexes with aggregated O is expected.

We want to remark that our model simulates the local environment of Er ions in oxygen co-doped GaN samples independently of the growth process. Energetic barriers between defect configurations, diffusivity of oxygen, erbium and intrinsic GaN defects, as well as other factors that can be lead back to the actual growth process do not enter the simulation. The results obtained refer therefore either to an "ideal" MBE-growth (i.e. high temperature and extremely low atomic fluence) or to an implantation process followed by an "ideal annealing" which completely removes any residual lattice damage. Our calculations reveal that different Er-O defect configurations are energetically very close. Therefore other "external" factors (i.e. the actual growth process) will determine which defect configuration will be formed in actual samples, rather than their relative formation energy.

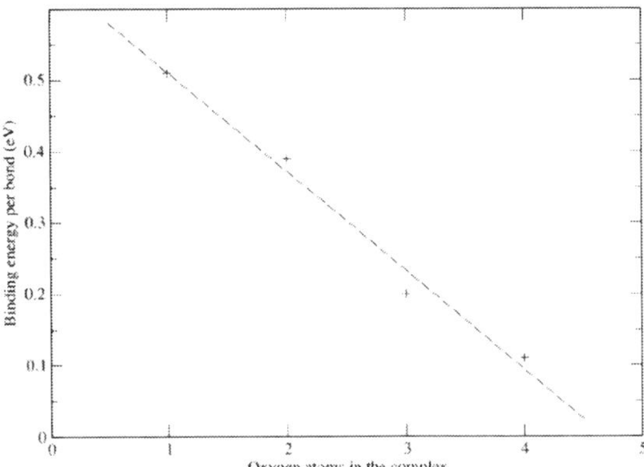

Figure 2. Cohesive energy *per bond* of the Er-O$_x$ complexes. The cohesive energy per bond decreases almost linearly with the number of oxygen atoms, indicating that in moderately doped material the most favorable configuration is the Er-O pair. The dashed line is a linear fit of the calculated values.

However, as the Er-O binding energy of about 0.5 eV is fairly high[1], it can be concluded that Er-O co-doping of otherwise undoped GaN will yield to the formation of Er-O complexes. The number of O-ions that are effectively bound to Er-ions after annealing can be estimated with a simple model. Labeling with c_{Er}, c_O and c_{Ga} the Er and O-concentration and the density of Ga lattice sites, an estimate of the percentage c_{Er-O} of bound Er and O ions is given by:

$$c_{Er-O} = \frac{c_{Er}c_O}{c_{Er} + c_{Ga}e^{-\frac{E}{kT}}}$$

where T is the temperature and k the Boltzmann constant. Considering c_{Er} to be as in our simulation and similarly to the real sample about 0.078%, we find after annealing at 1000 K an important fraction of the oxygen (about 30-40%) to be bound with the rare earth. The percentage of bound oxygen as a function of the annealing temperature is plotted in Fig. 3.

---

[1] Even if it is much smaller than in the case of Er-O co-doped Si.

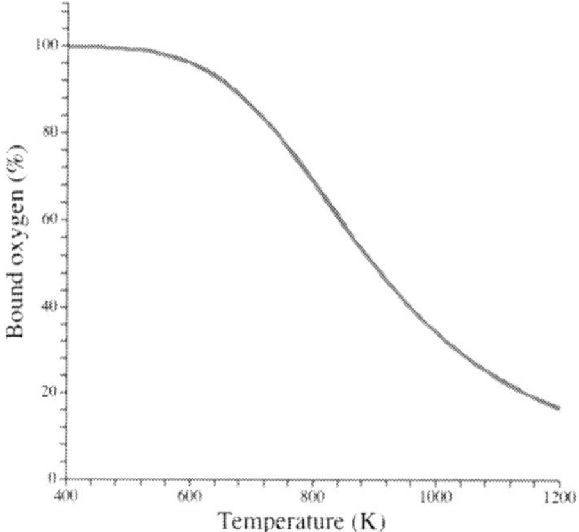

Figure 3. Fraction of oxygen ions bound to an erbium atom as a function of the annealing temperature.

CONCLUSIONS

We have presented a thorough theoretical investigation of the formation of Er-O complexes in hexagonal GaN. Oxygen atoms readily substitute N atoms introducing only a minor lattice strain in the host. This is probably due to the similarity (electronegativity and atomic radii) of the two ions, in particular in the case of O^+. The existence of defect configurations essential to the luminescence in other hosts, such as $Er-O_2$ and Er_2-O_3 could not be verified. Instead, several Er-O pairs of different symmetry are formed. Unfortunately our calculations do not directly reveal any details about the luminescence properties of the single defect configurations. It can be stated however, that all the individuated characteristics of the defect center are fully compatible with the present knowledge of the luminescent centers:

- Er-O pairs are bound and exist even after annealing.
- The crystal field distortion induced by the presence of oxygen will relax the selection rules.
- The Er valence state in this complex is 3+.
- The defect symmetry coincides to the "observed" C_{3v}. Indeed, the C_{1h} symmetry of the basal pairs will result in average in C_{3v} symmetry in experiments in which the contributions from different centers is added in the revealed signal (such as EXAFS).
- The presence of a shallow donor level might help to transfer energy to the f-shells.

Furthermore, the presence of different configurations of similar formation energy could explain the luminescent site multiplicity observed in co-doped samples [9,13,14]. Therefore, even if we cannot directly conclude that Er-O pairs are involved in the luminescence process, we cannot exclude it, as in the case of Er-O$_x$ complexes. The latter are, because their very high formation energy and very low binding energy, not likely to occur at all.

ACKNOWLEDGMENTS

The calculations were done using grants of computer time from the Paderborn Center for Parallel Computing (PC^2) and the Höchstleistungs-Rechenzentrum Stuttgart (HLRS). The Deutsche Forschungsgemeinschaft is acknowledged for financial support.

REFERENCES

1. D.S. Lee and A.J. Steckl, Appl. Phys. Lett. **81**, 2331 (2002).
2. *Rare Earth Doped III-Nitrides for Optoelectronic and Spintronic Applications*, K.P. O'Donnell and V. Dierolf editors (Springer, Dordrecht, 2010).
3. I.S. Roqan, K.P. O'Donnell, R.W. Martin, P.R. Edwards, S.F. Song, A. Vantomme, K. Lorenz, E. Alves and M. Bockowski, Phys. Rev. B **81**, 085209 (2010).
4. J. Michel, J.L. Benton, R.F. Ferrante, D.C. Jacobson, D.J. Eaglesham, E.A. Fitzgerald, Y.-H. Xie, J.M. Poate and L.C. Kimerling, J. Appl. Phys. **70**, 2672 (1991).
5. A. Nishikawa, T. Kawasaki, N. Furukawa, Y. Terai, and Y. Fujiwara, Appl. Phys. Lett. **97**, 051113 (2010).
6. M.Yoshida, K. Hiraka, H. Ohta, Y. Fujiwara, A. Koizumi and Y. Takeda, J. Appl. Phys. **96**, 4189 (2004); **97**, 023909 (2005).
7. M. Fujisawa, A. Asakura, F. Elmasry, S. Okubo, H. Ohta and Y. Fujiwara, J. Appl. Phys. **109**, 053910 (2011).
8. P.N. Favennec, H. L' Haridon, D. Moutonnet, M. Salvi and M. Gauneau, Jpn. J. Appl. Phys., Part 2 **29**, L524 (1990).
9. J.T. Torvik, C.H. Qiu, R.J. Feuerstein, J.I. Pankove and F. Namvar, Appl. Phys. Lett. **69**, 2098 (1996); J. Appl. Phys. **81**, 6343 (1997).
10. C.H. Qiu, M.W. Leksono, J.I. Pankove, J.T. Torvik, R.J. Feuerstein and F. Namvar, Appl. Phys. Lett. **66**, 562 (1995).
11. P.H. Citrin, P.A. Northrup, R. Birkhahn and A.J. Steckl, Appl. Phys. Lett. **76**, 2865 (2000).
12. E. Alves, T. Monteiro, J. Soares, L. Santos, M.F. da Silva, J.C. Soares, W. Lojkowski, D. Kolesnikov, R. Vianden and J.G. Correia, Mat. Sci. Engin. **B81**, 132 (2001).
13. T. Monteiro, J. Soares, M.R. Correia and E. Alves, J. Appl. Phys. **89**, 6183 (2001).
14. J.M. Zavada, M. Thaik, U. Hömmerich, J.D. McKenzie, C.R. Abernathy, S.J. Pearton and R.G. Wilson, J. Alloys and Compounds **300-301**, 207 (2000).
15. J.M. Zavada, C.J. Ellis, J.Y. Lin, H.X. Jiang, J.T. Seo, U. Hömmerich, M. Thaik, R.G. Wilson, P.A. Grundowski and R.D. Dupuis, Mat. Sci. Engin. **B81**, 127 (2001).
16. J.-S. Filhol, R. Jones, J. M. Shaw and P.R. Briddon, Appl. Phys. Lett. **84**, 2841 (2004).
17. K. Takahei, A. Taguchi, Y. Horikoshi and J. Nakata, J. Appl. Phys. **76**, 4332 (1994).

18. T. Ishiama, E. Katayama, K. Murakami, K. Takahei and A. Taguchi, J. Appl. Phys. **84**, 6782 (1998).
19. R. Sema, M. Lohmeier, P.M. Zagwijn and A. Polman, Appl. Phys. Lett. **66**, 1385 (1995).
20. A.G. Raffa and P. Ballone, Phys. Rev. B **65**, 121309(R) (2002).
21. G. Kresse, J. Furthmüller, Phys. Rev. B **54**, 11169 (1996).
22. J.P. Perdew, Y. Wang, Phys. Rev. B **33**, 8800 (1986).
23. P.E. Blöchl, Phys. Rev. B **50**, 17953 (1994).
24. S. Sanna, W.G. Schmidt, Th. Frauenheim and U. Gerstmann, Phys. Rev. B **80**, 104120 (2009).
25. S. Sanna, B. Hourahine, Th. Frauenheim and U. Gerstmann, Phys. Stat. Sol (c) **5**, 2358 (2008).
26. B. de Vries, V. Matias, A. Vantomme, U. Wahl, E.M.C. Rita, E. Alves, A.M.L. Lopes and J.G. Correia, App. Phys. Lett. **84**, 4304 (2004).
27. B. de Vries, *Lattice site location of impurities in group III nitrides using emission channeling*, Ph.D thesis, University of Leuven (2006).
28. R. Jones, Opt. Mat. **28**, 718 (2006) and references therein.
29. C.G. Van de Walle and J. Neugebauer, J. Appl. Phys. **95**, 3851 (2004).

Mater. Res. Soc. Symp. Proc. Vol. 1342 © 2011 Materials Research Society
DOI: 10.1557/opl.2011.999

Photoluminescence of Eu-doped GaN

K.P.O'Donnell, SUPA Dept. of Physics, University of Strathclyde, Glasgow G4 0NG, UK

ABSTRACT

This talk reviews work on the optical properties of Eu-doped GaN at the Semiconductor Spectroscopy laboratory of the University of Strathclyde. The principal experimental technique used has been lamp-based Photoluminescence/Excitation (PL/E) spectroscopy on samples produced mainly by high-energy ion implantation and annealing, either at low or high pressures of nitrogen, as described by Lorenz et al. [1]. These have been supplemented by samples doped in-situ either by Molecular Beam Epitaxy or Metallorganic Vapour Phase Epitaxy. Magneto-optic experiments on GaN:Eu were carried out in collaboration with the University of Bath.

INTRODUCTION

GaN doped with Europium, hereafter GaN:Eu, is currently of interest as a source of red light in prospective laser and light-emitting diodes. Despite quite a large recent effort, the defect structure, excitation mechanism and even the basic spectroscopy of GaN:Eu, in common with other rare-earth ion and III-nitride combinations, is not well understood.

EXPERIMENTAL DETAILS

Photoluminescence/Excitation (PL/E) spectroscopy at the University of Strathclyde employs a home-built apparatus (figure 1) which couples a 1 kW Xe arc lamp with a ¼ m Jobin-Yvon monochromator as the excitation source and a McPherson 2/3 m spectrometer with a cooled photomultiplier tube (PMT) for analysis and detection of the sample fluorescence. Samples, mounted in an optical cryostat on an x-z stage, are cooled by a 2-stage closed-cycle Helium refrigerator to a base temperature of ~15 K. An Oxford Intruments temperature controller uses a strip heater to elevate the sample temperature in the range 15 – 300 K.

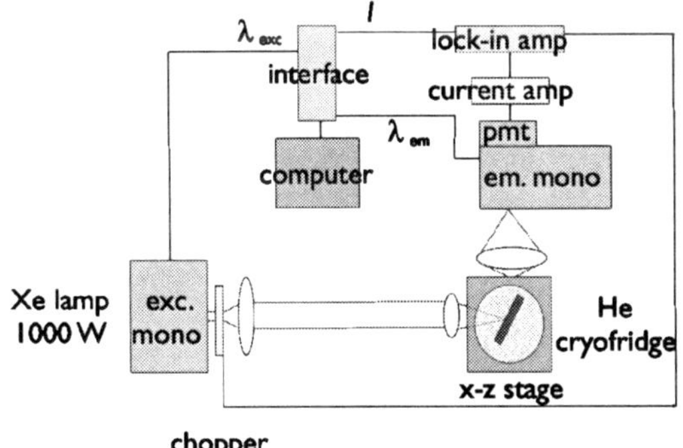

Figure 1: PL/E spectroscopy apparatus.

The PL/E system can be used with a fixed excitation wavelength to obtain photoluminescence (PL) spectra by recording the amplified PL signal as a function of the emission wavelength. Excitation spectra monitor the PL signal at a particular wavelength of emission while scanning the excitation spectrometer wavelength. A digital interface controls the monochromators by means of stepper motors and the spectra are recorded and displayed by computer.

The dispersion of the emission monochromator (0.6 nm per mm of slit with a 2400 line mm^{-1} grating) allows good resolution of sharp emission lines, which are typically ~0.2-3 nm wide (FWHM) at ~620 nm (i.e. < 0.5 meV), while the excitation monochromator is usually set to provide an adequate sample illumination of order 1 mW cm^{-2} in a 10 meV bandpass.

Preliminary room temperature cathodoluminescence (CL) spectroscopy complements the PL/E measurements. In general CL uses very much higher excitation densities than PL/E.

EXPERIMENTAL RESULTS AND DISCUSSION

MOCVD samples

By far the brightest GaN-based red-light emitters examined to date have been MOCVD-doped samples produced at Osaka University by Fujiwara and co-workers [2]. We begin with an overview of the cathodoluminescence (CL) and PL/E of these samples and later compare them with ion-implanted material.

Figure 2 shows a typical low-resolution RTCL spectrum in the wavelength range from 350 to 800 nm of a GaN:Eu sample grown by MOCVD at the near-optimum temperature of 1000 °C.

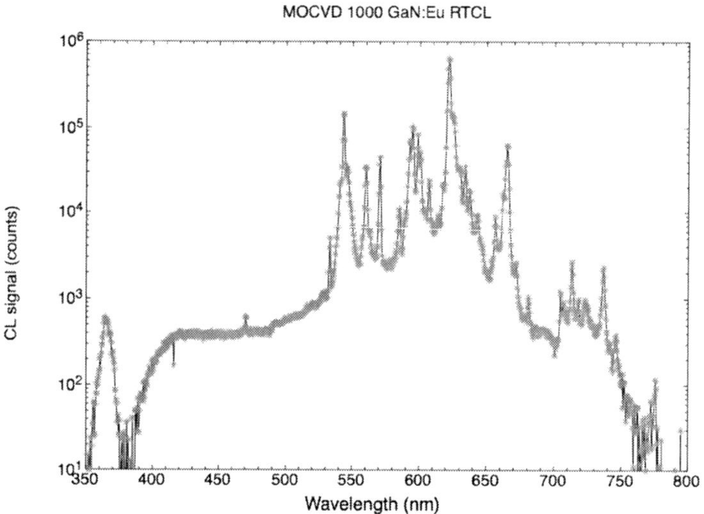

Figure 2: RTCL spectrum of GaN:Eu MOCVD sample.

The broad emission peak between 350 and 380 nm is due to intrinsic near-bandedge luminescence of the host material, while the sharp emission lines between about 550 and 770 nm can all be assigned to Europium ions in the 3+ charge state. The sharp lines represent intra-*4f* shell transitions, forbidden and therefore latent in the free rare earth (RE) ion, but expressed or developed when the ion is incorporated in a solid host, through the orbital intermixing by the crystal field first described separately by Judd and Ofelt [3].

Much more spectral data is provided by PL measurements carried out at low temperatures, as shown in figure 3. The use of two different excitation wavelengths, chosen to be above (350 nm) and below (390 nm) the GaN band edge, allows a useful discrimination to be made between different emitting centres with overlapping emission lines [4]. Thus the terms 'indirect' and 'direct' excitation broadly distinguish mechanisms whereby the RE ion is excited via electron-hole pair migration or otherwise.

In figure 3, the approximate wavelength locations of certain transitions are indicated using term symbols of the usual Russell-Saunders form $^{2S+1}L_J$ where S, L and J represent spin, orbital and total angular momentum quantum numbers, respectively. The strongest and best-studied transition is $^5D_0 \rightarrow {}^7F_2$ near 620 nm which is split into as many as $2J + 1 = 5$ components, depending upon the site symmetry, by the crystal field.

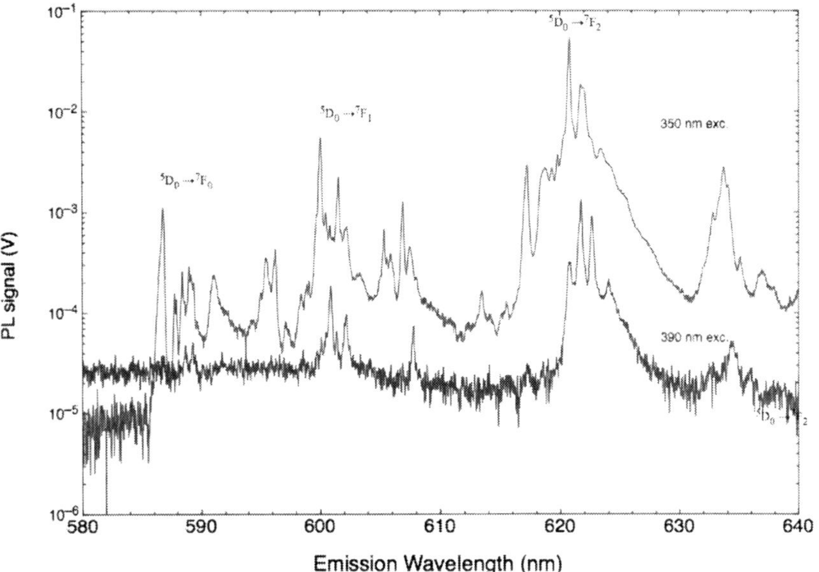

Figure 3: Red-orange PL at 20 K of GaN:Eu MOCVD sample

The general profile of a transition feature comprises a multiplet of several lines superimposed upon a broad background. (Note that the spectra presented in figures 2 and 3 are in semi-logarithmic form, which exaggerates somewhat the background contributions.)

From previous work on ion-implanted GaN:Eu [4] we have come to understand that the sharp-line PL originates mainly on 2 well-defined defect centres, called Eu1 and Eu2. Eu2 is hard to excite with below-gap light, while Eu1 has a characteristic below-gap excitation band at ~380 nm, called X1 [4]. High-resolution spectra in linear format, figure 4, clearly reveal which lines belong to each centre (as can also be determined from figure 3) but also show overlaps near 621 nm and 622 nm. However the Eu2 contribution at the first of these wavelengths is estimated to be about 100 times that of Eu1; hence we chose emission wavelengths of ~621 nm and ~623 nm to monitor Eu2 and Eu1, respectively, in our excitation studies.

The PL/E spectra of Eu1 and Eu2 centres are compared in figure 5. Both feature a dominant above-gap component (λ_{ex} < 354 nm) corresponding to excitation of RE emission through the generation of free electron-hole pairs. There is a weak indication of a resonant enhancement of the Eu2 centre just above the band edge but the Eu1 PL/E maximum occurs at somewhat higher energy. The spectra have not been corrected for monochromator throughput nor lamp output which rises monotonically with wavelength in the region of interest. The Eu1 PLE shows a distinct cusp at 362 nm which we take to indicate the limit of the 'Urbach tail'.

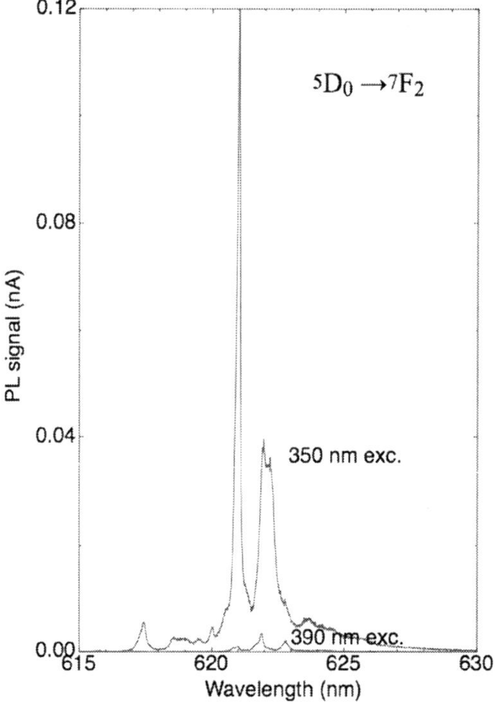

Figure 4: PL spectrum of MOCVD GaN:Eu under (a) 350 nm and (b) 390 nm excitation. The contribution of Eu1 to the above-gap spectrum is clearly rather small for this sample.

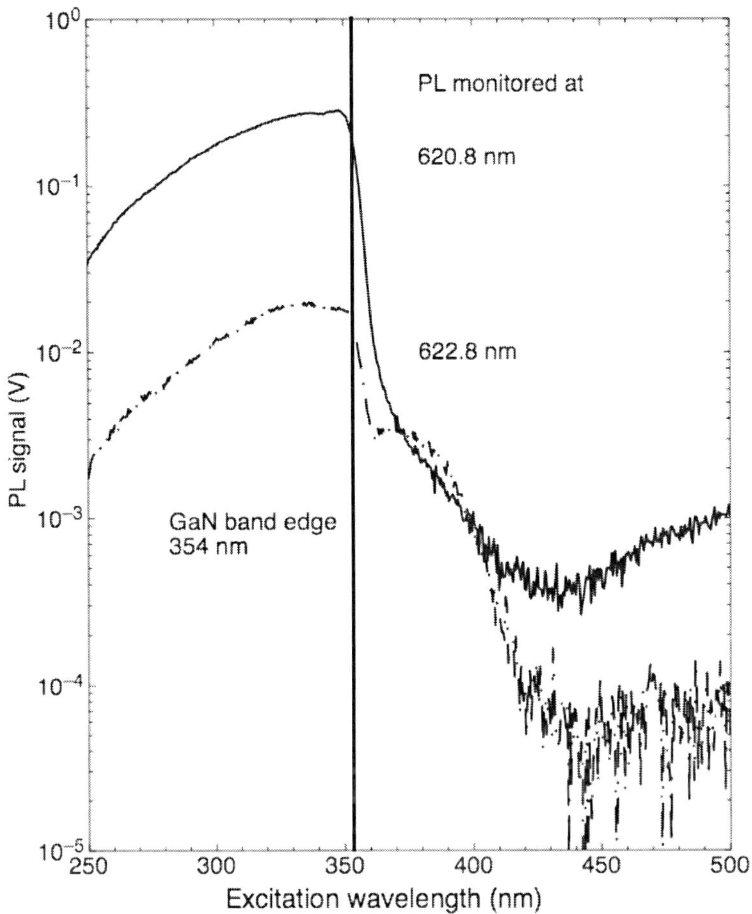

Figure 5: PLE spectra of MOCVD GaN:Eu sample emission monitored at (a) 620.8 nm, corresponding to Eu2, and (b) 622.8 nm, corresponding to Eu1. The GaN bandedge at 354 nm should separate 'indirect' from 'direct' excitation mechanisms, but see text.

Hence RE excitation in the range 354 nm (3.50 eV) to 362 nm (3.43 eV) in GaN involves tunneling of electrons and/or holes assisted by spatial disorder. Between 362 nm and ~425 nm (~2.92 eV) the X1 band produces a distinct excitation feature for Eu1, the origin of which is still controversial, but which may involve bound exciton or charge-transfer states intermediate between Eu^{3+} and Eu^{2+} [4]. The defect Eu2 can also be excited via X1, but with an efficiency of order a hundred times smaller than that of the band edge mechanism. In addition an undetermined fraction of the excitation shoulder in this region (for 620.8 nm detection) can be ascribed to the presence of components of Eu1 which overlap at this wavelength (see figures 4,7).

Above 425 nm, the situation regarding excitation becomes a little more complicated [5]. We provide here only a brief summary of the lamp-based (low power) results. Firstly, we fail to detect direct excitation signals involving the internal multielectron states of Eu^{3+}. This failure is probably due to the low absorption oscillator strengths ($f \sim 10^{-4}$) expected for intra-$4f$ transitions, which makes the ions difficult to excite directly. Instead of sharp lines, we do observe one or two indefinite broad bands, particularly when detecting Eu2 emission. The rising edge of the main band, with a peak near 530 nm, can be seen in Figure 5, as well as an indication of a second, weaker, band peaking at 475 nm. The origin of these bands is unclear at present, but we assume that they involve defects.

In order to quantify these observations, we define two experimentally derived parameters based upon the PL/E signal strength at the *minimum* of the excitation spectrum. This minimum is found between 425 and 440 nm, depending on the monitored transition and, to a lesser extent, on the sample. The first parameter, referred to as the sample 'brightness' is the ratio:

B = PLE(max)/PLE(min).

For the spectra shown in Figure 5, B ~ 800 for the 620.8 nm line and ~500 for the 622.8 nm line. The higher the value of B, the cleaner will be the PL/E spectra. The second parameter refers to the effects of defects on the sample excitation:

D = PLE (530 nm)/ PLE (min)

D is very sensitive to the monitored transition, being generally ~1 (no defect enhancement) for Eu1 and of order 3-10 for Eu2, depending on the sample. D = 1 is the minimum value possible while a sample with a high value of D would be considered to be 'dirty'. These parameters give a crude indication of sample 'quality' and may be used to quantify the effects of different preparation techniques.

Ion-implanted GaN:Eu

Compared to MOCVD samples, ion-implanted material presents somewhat weaker PL spectra, which may however represent a much larger usability of the incorporated ions [6]. Figure 6(a) compares typical spectra of as-grown and high-pressure (HP)-annealed, implanted GaN:Eu. (HP annealing is carried out at pressures of order 10^4 bar of N_2, whereas low-pressure (LP) annealing uses typically 'only' a few bar. [1]) Figure 6(b) compares PL spectra of HP and LP samples that were implanted under identical conditions.

While annealing at HP increases the PL efficiency by a factor of 30-50, compared to LP annealing at the highest possible temperatures, already several hundred K above the growth temperature, the MOCVD samples are typically 10-20 times brighter again.

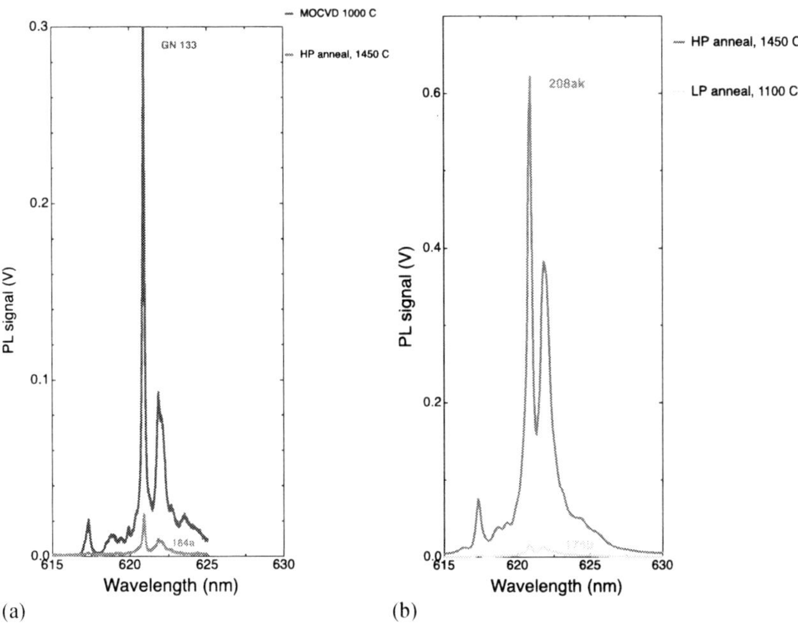

Figure 6(a) compares PL from MOCVD and HP-annealed GaN:Eu samples, while 6(b) compares HP with LP annealing.

An experiment in which an MOCVD sample was annealed under HP at a temperature of 1450 °C, successfully increased the red luminescence output by a factor of 3.3. Since this increase was accompanied by a decrease in the background luminescence referred to earlier, it represents a significant increase of the Eu2 defect population of MOCVD samples brought about by annealing.

GaN:Eu PL polarisation and magneto-optic spectroscopy

Both Eu1 and Eu2 involve Eu atoms that are substitutional on the Ga site in wurtzite GaN [7]. They are therefore expected to show polarized luminescence in agreement with well-known selection rules. Figure 7 shows how polarized detection affects the $^5D_0$ to $^7F_2$ PL spectrum of the Eu1 defect, preferentially excited in the X1 band at 390 nm. The polarized spectra are somewhat sharper than those obtained with unpolarised light, revealing the maximum five-fold splitting of the $^7F_2$ level by a low-symmetry crystal field. These results will be published in detail elsewhere.

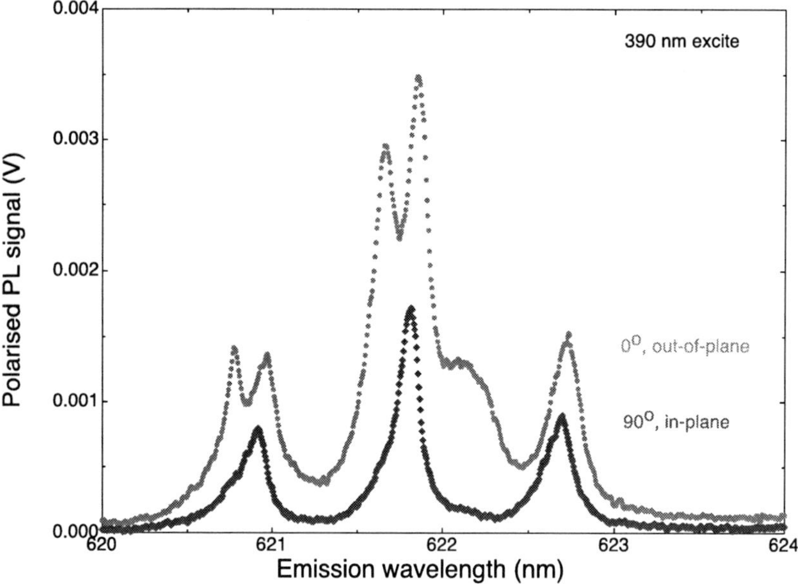

Figure 7: PL spectra detected with a linear polarizer placed between the sample and the detection monochromator. The sample is excited along the c-axis and the light emission from the sample edge is analysed.

Magneto-optic studies of HP-annealed implanted samples were carried out in collaboration with the University of Bath [8]. These studies highlight, once again, the complex nature of Eu^{3+} optically active centers in GaN. Not only do emission lines for Eu1 and Eu2 centres split differently in a magnetic field parallel to the c-axis, but emission lines belonging to the same center also split differently with g-factors for different lines ranging from ~ 1 to ~3. Moreover, the lines due to minority centres have g-factors ranging from 0 to ~6. The observed Zeeman splittings are associated with magnetic moments of the final $^7F_2$ state of the $^5D_0$-$^7F_2$ transition.

CONCLUSIONS

While much work has been done on the spectroscopy of Eu-doped GaN, much remains to be done, aided by the general availability of better samples and improved understanding of the defect configurations and excitation mechanisms. In parallel with this, the further development of the GaN:Eu diode will continue to drive the field forward.

ACKNOWLEDGMENTS

I would like to take the opportunity to thank my many collaborators and co-authors, some of them ex-students, especially those named in the references below, and in fact all of the contributors to the book named in Refs [1] and [4].

REFERENCES

[1] K Lorenz, E Alves, F Gloux, P Ruterana in *Rare-Earth Doped III-Nitrides for Optoelectronic and Spintronic Applications* (Kevin O'Donnell and Volkmar Dierolf, Eds; Springer, 2010), Ch2
[2] Atsushi Nishikawa, Takashi Kawasaki, Naoki Furukawa, Yoshikazu Terai, and Yasufumi Fujiwara, Applied Physics Express 2 (2009) 071004
[3] B R Judd, Phys. Rev. 127, 750 (1962); G S Ofelt, J. Chem. Phys. 37, 520 (1962)
[4] R W Martin in *Rare-Earth Doped III-Nitrides for Optoelectronic and Spintronic Applications* (Kevin O'Donnell and Volkmar Dierolf, Eds; Springer, 2010), Ch7 and refs therein
[5] K.P. O'Donnell et al., Opt. Mater. (2010), doi:10.1016/j.optmat.2010.07.002
[6] K Lorenz, E Alves, I. S Roqan, K. P O'Donnell, A Nishikawa, Y. Fujiwara, M Bockowski, Appl. Phys Lett., 97, 111911 (2010)
[7] I. S Roqan, K. P. O'Donnell, R. W Martin, P. R Edwards, S. F Song, A Vantomme, K Lorenz, E Alves, M Bockowski, Phys. Rev. B81, 085209 (2010)
[8] V. Kachkanov, K.P. O'Donnell, C. Rice, D. Wolverson, R.W. Martin, K. Lorenz, E. Alves, M. Bockowski, MRS Fall Meeting 2010, to be published.

Mater. Res. Soc. Symp. Proc. Vol. 1342 © 2011 Materials Research Society
DOI: 10.1557/opl.2011.1156

Site Selective Magneto-Optical Studies of Eu ions in Gallium Nitride

Nathaniel Woodward[1], Atsushi Nishikawa[2], Yasufumi Fujiwara[2], and Volkmar Dierolf[1]

[1]Physics, Lehigh University, Bethlehem, Pennsylvania, U.S.A.

[2]Division of Materials and Manufacturing Science, Osaka University, Osaka, Japan.

ABSTRACT

We report site-selective studies of the Zeeman splittings that are observed for magnetic fields up to 6.6T for different Eu incorporation sites in GaN. Utilizing resonant excitation with visible light, we are able to distinguish the site and find for one center (Eu1) a splitting into five components as expected for C_{3v} symmetry. The corresponding g-values are 1.66 and 1.90. The two lines of another center Eu2 each split into two levels corresponding to g-values of 1.9 and 2.84. Most surprisingly a third center, for which only one line is clearly identified, a g-value of 6.16 is found which is larger than can be explained for a $^7F_2$ purely ionic Eu state.

INTRODUCTION

Eu ions in GaN have been the topic of intense spectroscopic studies due to their application for electroluminescence devices. Several different incorporation sites have been identified by different groups [1-4]. Most notable, two centers Eu1 and Eu2 have been studied for their excitation efficiency under above band-gap excitation and their relative numbers in samples that have been grown or thermally annealed in a different ways. In order to further characterize the spectroscopic properties of these centers, measurements under the application of magnetic fields are an excellent tool to test models such as for instance crystal field analysis, which could yield important information about the site symmetry. Measurement of the Zeeman splitting of optical spectra is especially important for Eu^{3+} for which no ground state EPR data can be obtained due to the ^7F_o character of its ground state. Kachkanov et al. have reported first results [5], which only offered limited site-selectivity and hence are difficult to interpret. The purpose of this report is to provide site-selective Zeeman data under obtained under resonant excitation conditions, which yield the g-factors of all sublevels of the $^7F_2$ state.

The two samples used in our experiment were grown by OMVPE at 1050° C under 10 kPa of pressure. For more information on the growth of these samples refer to [6]. The detailed spectroscopic data were reported in Ref. 1 and 2.

RESULTS

For all of the magnetic measurements, we utilize Faraday geometry, where the polarization of the incident laser is perpendicular to the applied magnetic field. The emission is collected in backscattering geometry. We use a fiber-based optical access module, which is placed into a magnetic cryostat (manufacturer: Janis model: 9T Cryostat). As a fiber-coupled tunable laser source we use an Argon ion laser pumped dye laser (manufacturer: Coherent, model 590) that is tunable from 565nm to 600nm. For our Zeeman studies, we focus on the spectral range around 570nm in which the phonon-assisted $^7F_0$ to $^5D_0$ excitation transition are located. As shown in Ref. [2] this give the strongest excitation peaks and good site selectivity. In Fig. 1, we present combined excitation emission data for this spectral region for both without and with a field of 5T. These data allow us to identify several sites in terms of their excitation energies. Due to

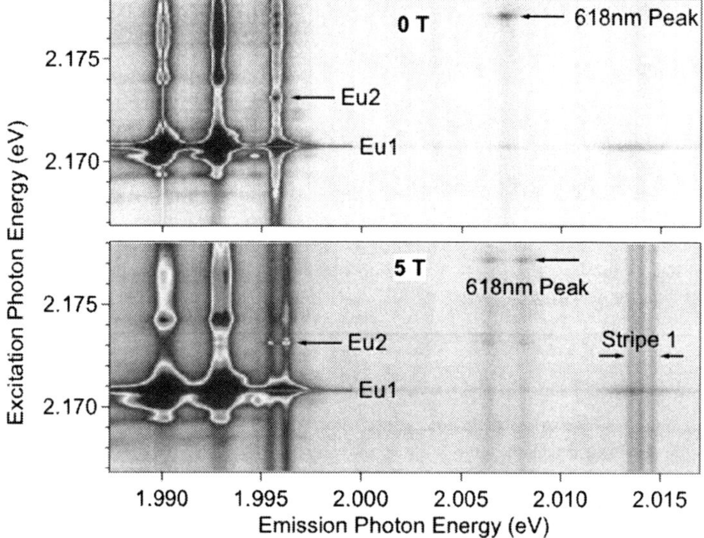

Fig. 1: Combined excitation emission data for the phonon-assisted $^7F_0$ to $^5D_0$ resonant excitation transition at around 570nm without and with a field of 5T applied in growth direction. The resonant excitation transition for the two most apparent sites Eu1 and Eu2 are indicated along with an additional peak that occurs at around 618nm in emission. For the data at 5 T an additional "stripe" feature becomes apparent.

the J=0 nature of the ground and excited state the excitation position for these centers is not changed and hence a constant excitation energy can be used to follow the Zeeman splitting for each site as the applied magnetic field is changed. The Eu2 center can also be excited non-resonantly in this spectral regime as indicated by the vertical stripes that extend throughout the image. We also note that the strength of the vertical stripes increase with magnetic field suggesting that this excitation channels get enhanced with magnetic field.

In Fig. 2, individual spectra that are obtained under the described selective excitation conditions are shown for the Eu1 and Eu2 center. The Eu1 center has, as expected for C_{3v} symmetry, three transitions that split into five under the applied field. Due to above mentioned non-resonant excitation of the Eu2 center, we still see a small

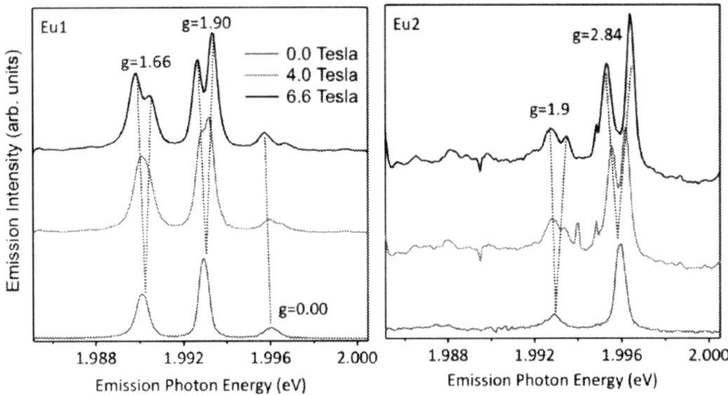

Fig. 2: Individual emission spectra for different magnetic field values obtained by selectively exciting Eu1 (left) and Eu2 (right). The resulting effective g-values are indicated.

perturbation from this center remains for the high-energy transition. The two lines of the Eu2 center that have been observed each split into two lines in a magnetic field. The splitting can be converted to effective g-factors using $\Delta E = \mu_b g_{eff} B$. The effective g-values of the $^7F_2$ can thus be determined by solving for g_{eff}, or $g_{eff} = \Delta E / \mu_b B$, where $\mu_b = 5.788 \times 10^{-5} eV \cdot T^{-1}$ is the Bohr magneton. We find for the two centers effective g-factors of 0, 1.90, 1.66 and 2.84 and 1.9, respectively.

In Fig. 3, we show individual spectra for resonant excitation of the 618nm peak. Its splitting is considerably more pronounced than that of the other peaks. It corresponds to an effective g-factor of 6.16. In this figure, the splittings of two stripe-like features in our CEES data are also presented. For these features, a single peak splits into three components suggesting higher site symmetry than C_{3v}.

DISCUSSION AND CONCLUSIONS

For Eu ions in a C_{3v} symmetry site we expect two Kramers doublets, which split in a magnet field, and singlet that does not change. This is what we observe for both the Eu1 and the Eu2 (if we assume a third line that is not observed in our geometry). The effective g-factors are determined as splitting factors, which result from the perturbation Hamiltonian matrix elements using the states that have been split by the crystal field. They are best determined in a crystal field analysis. To estimate the range of values that are possible, we can use LS coupling notation and group theory arguments. The levels can then be described by crystal field quantum number and are linear combination of levels with M_J values of -2,-1,0,1, and 2 for the $^7F_2$ state. The effective g-factors are then determined by $g_J (|a_1| M_{J1} + |a_2| M_{J2})$ where a_i are the mixing coefficients, g_J is the Lande g-factor of 1.5, and M_{J1}, M_{J2} are -2, 1 or -1, 2 based on the C_{3v} symmetry of the Eu ions. The effective values observed for Eu1 and Eu2 can hence be explained in this model.

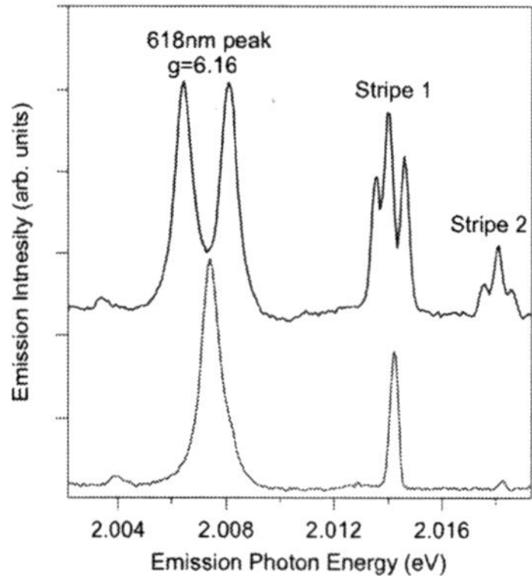

Fig. 3: Individual emission spectra for different magnetic field values obtained by selectively exciting of the 618nm peak The resulting effective g-value is indicated.

Crystal field analysis that combines both the crystal field splitting and the g-factors can bring more inside into the symmetry and potential perturbation of the centers.

On the other hand, the splitting of the 618nm line is unusually large and cannot be accounted for within the simple picture and suggest that levels with higher J values or admixture from the host levels is involved. Other observations about this line suggest it may be connected with an acceptor state.

Overall, our studies provided new insight into the electronic state of the Eu ion in GaN. Nevertheless, more studies and modeling are needed to explain the observed behavior and to exploit their quantitative aspects.

ACKNOWLEDGMENTS

This work was supported by a grant from the National Science Foundation Grant No. NSF-DMR 0705217. Y.F. and A.N. would like to acknowledge gratefully financial support by Grant-in-Aid for Creative Scientific Research Grant No. 19GS1209 from the Japan Society for the Promotion of Science, and by Grant-in-Aid for Young Scientists Grant No. 21760007 and by the Global Centre of Excellence Program 'Advanced Structural and Functional Materials Design' from the Ministry of Education, Culture, Sports, Science and Technology of Japan.

REFERENCES:

1. N. Woodward, J. Poplawsky, B. Mitchell, A. Nishikawa, Y. Fujiwara, and V. Dierolf, Appl. Phys. Lett. **98**, 011102 (2011).
2. N. Woodward, A. Nishikawa, Y. Fujiwara, and V. Dierolf, Opt. Mater. **33**, 7 (2010).
3. Z. Fleischman, C. Munasinghe, A. J. Steckl, A. Wakahara, J. Zavada, and V. Dierolf, Appl. Phys. B: Lasers Opt. **97**, 607 (2009).

4. I. S. Roqan, K. P. O'Donnell, R. W. Martin, P. R. Edwards, S. F. Song, A. Vantomme, K. Lorenz, E. Alves, and M. Bockowski, Phys. Rev. B **81**, 085209 (2010).

5. V. Kachkanov, K.P. O'Donnell, C. Rice, D. Wolverson, R.W. Martin, K. Lorenz, E. Alves and M. Bockowski, MRS Online Proceedings Library / Volume 1290 / mrsf10-1290-i03-06 (Proceedings MRS Fall Meeting 2010).

6. A. Nishikawa, T. Kawasaki, N. Furukawa, Y. Terai, and Y. Fujiwara, Appl. Phys. Express 2, 071004 (2009).

AUTHOR INDEX

CPSIA information can be obtained at www.ICGtesting.com
Printed in the USA
BVOW040746081111

275345BV00003B/1/P